LIGHT AND LIFE

Light and Life

Michael Gross

OXFORD
UNIVERSITY PRESS

OXFORD
UNIVERSITY PRESS

Great Clarendon Street, Oxford OX2 6DP

Oxford University Press is a department of the University of Oxford.
It furthers the University's objective of excellence in research, scholarship,
and education by publishing worldwide in

Oxford New York

Auckland Bangkok Buenos Aires Cape Town Chennai
Dar es Salaam Delhi Hong Kong Istanbul Karachi Kolkata
Kuala Lumpur Madrid Melbourne Mexico City Mumbai Nairobi
São Paulo Singapore Taipei Tokyo Toronto

Oxford is a registered trade mark of Oxford University Press
in the UK and in certain other countries

Published in the United States
by Oxford University Press Inc., New York

British Library Cataloguing in Publication Data
Data available

Library of Congress Cataloging in Publication Data
Data available

ISBN 0-19-856480-5

10 9 8 7 6 5 4 3 2 1

Typeset in Poppl-Pontiflex
by George Hammond Design
Printed in Great Britain by
T. J. International Limited, Padstow, Cornwall

Contents

Confessions of a light-addict

LIGHT, like no other physical phenomenon, is linked in a wide variety of ways with the biological phenomenon of *life*. You are able to read this page, for instance, because light is reflected from it and carries the information to your retina, where it can be converted into nerve signals. While you're doing this, you breathe oxygen which was produced by green plants using the energy of sunlight. Your inner organs are breaking down food, the chemical energy of which can also be traced back to photosynthesis. You're sufficiently awake to understand these thoughts (unless you happen to be jet-lagged), because your inbuilt biological clock has set itself using cues of light and dark and tells you that you're still within the active period of your day. Thus, at this very moment, you are making direct or indirect use of three fundamentally important interfaces between light and life: vision, photosynthesis, and the biological clock.

This close relationship with light is something we share with most living species. There may be other kinds of life in the Universe, but the only kind that we know about owes its existence to the energy of the Sun, and its family tree was shaped by the invention of photosynthesis more than by any other event. As a result, today's inhabitants of our planet tend to rely on light for energy, information, and guidance in space-time. No wonder ancient cultures acknowledged its importance by worshipping the Sun as a life-giving, god-like being. Had this small yellow star acquired less fuel and failed to light up, all lifeforms that we know of would be non-existent. This would probably also be the result if you took the planet a bit further away from the Sun. Planet Earth orbiting between Mars and Jupiter would be as lifeless as the Moon, as I will show in the first chapter.

The second most important factor in shaping life on our planet involves the technological revolution which introduced a better way of making use of solar energy, but incidentally changed everything else too. If evolution hadn't come up with photosynthesis some 3.5 billion years ago, there would be no oxygen in the atmosphere, no protective ozone layer above it, and probably no life on dry land. Planet Earth minus the invention of photosynthesis would be inhospitable for all higher organisms that are around today, and would be peopled only by primitive bacteria in the oceans or under the surface of the Earth.

Unravelling photosynthesis was one of the major challenges of modern biology. From the fundamental studies tracing the paths of carbon atoms using radioactive isotopes, through to today's structure-function analysis helped by crystal structures of essential parts and even of one entire photosynthetic reaction centre, the quest to understand the natural solar powerstations has a rich history of difficulties. Similarly, mimicking their performance in technological systems is a goal which we only approach at a painstakingly slow pace. All these aspects of photosynthesis will be explored in the second chapter.

In contrast to the fundamental importance of sunlight and photosynthesis, the second connection between light and life, bioluminescence, is almost an eccentricity and adds an element of comic relief (maybe just what you need by the time you have reached Chapter 3). Although many organisms have obvious evolutionary reasons for producing light of their own, some cases are less easily explained, and there may even be organisms for which light is the unwanted byproduct of a biochemical reaction run for entirely unrelated reasons. Ironically, however, technological applications of bioluminescence systems are much more advanced than any technology derived from photosynthesis.

That organisms rely on light in their orientation in space and time is easily explained as a direct consequence of both the importance of light for the energy balance and the regularity with which diurnal and annual patterns of light change are repeated. More light often means more food, even for organisms not directly involved in photosynthesis. Movement towards light is therefore observed even in the most primitive groups of bacteria. And plants, of course, move towards the light for obvious reasons.

Plants are also interesting to study for the influence of light on biological time-keeping. As their 'behaviour' is more restricted than that of animals of similar complexity, elementary reactions triggered by changes in luminosity, day length, etc. can best be studied in plants. Only recently, breakthrough discoveries have been made which put the molecular cogwheels of plant circadian clocks in evolutionary relationships both with their counterparts in animals, including insects and mammals, and with the mechanisms of light-induced plant movement.

Animals like ourselves tend to have an inbuilt clock which can only crudely keep a 24-hour day and needs recalibration with the help of the light/dark changes every day. The way we are influenced by light changes is very much linked to ancient, subconscious processing pathways, including hormonal response. Maladjustment of biological rhythm or light intake can therefore make us ill and depressive without us even realizing that light is the underlying problem. Recent discoveries suggest that this response to light and dark may be independent of the eyes (see Chapter 4).

More conscious but still not entirely under our control is the process of seeing and perceiving. Although we think we know what we're looking at, psychological research has shown that perception is far from being a video recording of what we choose to look at. The cooperation between our rather ancient eyes (which come from the same line of development as those of all other vertebrates) and the much more recent conscious brain, with its higher-order processing and its mechanisms designed to make sense of the information flooding in, can come up with the most intriguing phenomena, many of which can be demonstrated by simple optical illusions (see Chapter 5).

In summary, *life* wouldn't be here without *light*. It wouldn't be as highly evolved as it is if it had not made the best possible use of light's energy and information content for using photosynthesis, biological clocks, and vision. It has also developed ingenious ways of producing its own light for a variety of purposes. Much of this insight results from scientific research conducted in the twentieth century. On the other hand, many of the earlier civilizations had a very special relationship to the Sun and its light, often attributing god-like qualities to it. In the last chapter I try to explore ways in which this pre-scientific worship of the Sun can be linked to what we now know about the relationship between light and life. More generally, I also investigate how the views on the subjects discussed here have changed through the ages.

This is, in a nutshell, what this book is about. Why did I write it? Unlike my two previous books, this one has no direct links to my own scientific research. Some of the more biochemical aspects of this subject are still in the neighbourhood of my expertise as a researcher, inasmuch as they deal with proteins, as I have done throughout my career. But to most of the subjects presented here I have come with the naïve curiosity of a science journalist who finds many things new and exciting and tries to communicate this excitement to the general public.

Although I am not claiming this to be a qualification, I myself am quite addicted to sunlight (as will become apparent in various places throughout the book), so this book is to a certain extent reflecting my own nature as much as nature in general. I have never been able to comprehend how people can work in offices with no natural light, or drive cars with an irremovable lid on for that matter.

And, if reasons are needed, here is one final reason for writing this book: I used to be intrigued as a child by the mathematics of cone sections. You take a double cone for instance, cut a plane through it at an oblique angle, and end up with some weird curves like parabolas or hyperbolas. As a writer I like to cut through the entire body of scientific knowledge at strange angles to reveal inter- esting sections. Light is, I hope, one angle that allows us to expose meaningful connections between apparently distant regions of science, ranging from the Big Bang to the future of our planet, from microbiology to astronomy, and from the first cell to the theory of consciousness.

Much of this might have remained beyond my comprehension

had it not been for the helpful hints and explanations of experts whom I consulted. I am very grateful to the *Professores* and *Doctores* Josephine Arendt, John Baines, David Leopold, Harald Paulsen, and Petra Stoerig for shedding some light on various aspects of my topic. A big thankyou to my most enduring guinea-pig reader, Kevin Plaxco, who has given valuable feedback on most parts of the manuscript. The final version has also benefited from constructive criticism offered by two anonymous referees acting for Oxford University Press, for which I am also grateful.

I wrote this book during the last two of my eight and a half years at the Oxford Centre for Molecular Sciences. I am particularly grateful to Chris Dobson, who allowed me to stay on after my research funding expired and I switched my activity to writing full-time. Some spotlights on Oxford landmarks scattered in for light entertainment could also be read as a farewell to Oxford University, as I am now moving my professional base camp to Birkbeck College, London. I was very pleased to find that they have a nice little roof terrace where I can get my daily fix of photons.

MICHAEL GROSS
Oxford, October 2001

CHAPTER 1

The right place at the right time

ONE OF THE FAVOURITE WALKS I tend to take on Saturday afternoons with my children leads us across the green fields and into the Science Area of Oxford University. We end up in front of a lawn with fresh dinosaur footprints. No, I'm not joking. At this very moment, as I'm writing these lines, they are only a couple of weeks old. But I'll admit that they are fakes. They are in fact concrete casts of the petrified traces that a lorry-sized, carnivorous *Megalosaurus* left in the northern parts of Oxfordshire some 168 million years ago. I totally admire the idea to put the casts there, even though it makes the sign saying 'no games allowed' look a little bit ridiculous. It's just so typical—letting the dinosaurs run free but banning the kids from playing on the grass. But what did you expect? After all, the dinosaur-haunted lawn is the front garden of the superb Victorian building which was the first science department at Oxford University—the University Museum of Natural History.

The museum was built between 1855 and 1860 as a three-winged neogothic structure, reminiscent of the French medieval cloth houses. The fourth wing was left out on purpose to allow for a possible extension, which was indeed built later on and now houses the anthropological treasure troves of the Pitt Rivers Museum. The newly built University Museum became the arena of a major piece of science history less than a year after its opening. It was here that the British Association for the Advancement of Science debated Darwin's theory of evolution in June 1860. A book entitled *On the origin of species* had just come out in November 1859 and was causing a bit of a stir. Darwin's disciples, Thomas Huxley and Joseph Hooker, apparently won the battle of words against the bishop of Oxford, Samuel Wilberforce, in front of some 700 people. Science had for the first time in history found a home in Oxford, and with that it also found a new degree of confidence in the debates with religious opponents.

The building also features prominently among the places where Lewis Carroll found inspiration for his *Alice* books. Most famously, the museum's dodo specimen, now derelict, found its way from these premises into the pages of *Alice in wonderland*. Twentieth-century residents of the museum include the Nobel prizewinning crystallographer Dorothy Crowfoot Hodgkin and the biologist and author Richard Dawkins.

Before we walk in, though, let us have a look at the impressive facade of the museum's front wing, with its dormers and first-floor windows arranged in immaculate symmetry around the central tower, while the ground-floor windows seem to be obeying more subtle rules, possibly derived from chaos theory. It only adds to the confusion that the carving of the window jambs has remained unfinished, after a dispute between the builders and architects concerning the finances of the work. Twenty metres to the right of the front wing, and connected by a narrow passage, we see a chapel-like structure known as the Abbot's Kitchen (with reference to the building at Glastonbury that served as its model). Built as a chemical laboratory that was part of the original museum plan, it related to the main building like the loo at the end of the garden related to the Victorian terrace house. In order to prevent any noxious smells from entering the main museum building, the planners banished chemistry to this ghetto. Maybe they were worried about explosions and fires too. Little did they know that this little appendage would proliferate into half a dozen major buildings over time. In a sense the museum can be called the mother of all science departments at Oxford. As new disciplines have come of age, they have one by one left this cosy nest and set up their own homes along South Parks Road.

Walking in through the heavy oak door feels very much like entering an old church, and even the noticeboard in the porch would still be compatible with this impression. Having scrambled up the stairs and entered a second, equally heavy oak door, one finally realizes that it's not a church, because churches don't normally have big skeletons in them. After a few more steps straight ahead, we find ourselves in a square courtyard under a glass roof supported by ornamented cast-iron pillars. They are thickly painted in a stone-like colour, but when you knock on them, the sound gives the metal core away. The whole structure is strangely reminiscent of a well-preserved early railway station, although the hushed atmosphere and the heated floor contradict this impression.

This courtyard, enclosed by the three main wings of the original building and by the Pitt Rivers Museum at the back, contains the larger exhibits and showcases, while smaller items are also displayed under the arcades around it. The items on show look very traditional at first glance—of course there is the usual mix of dinosaur skeletons, amethysts, and stuffed birds you expect to see in such a place, but there are also a few jewels to be discovered. Walking round the quadrangle you may notice that the columns supporting the arcades are all made from different kinds of stone—a geology course incorporated into the very substance of the building. And if you let your gaze follow the iron pillars towards the glass roof, you will notice that they have leaves at the top, like a forest of metal trees growing towards the light.

Coming back to the front wing of the quad, we walk up the staircase on the right-hand side corner (seen from the entrance). At the turn of the stairs we pass the museum's very own bee colony, buzzing behind perspex windows. On the first floor, we turn right and walk down along the right wing, passing millions of insects of all shapes and sizes, until we bump into the Snow-White-style glass shrine of a Roman inhabitant of the Science Area who probably would not have predicted that people would still stare at his skeleton some 1800 years after his death. Here we turn left into the back arcades, but stop short after a third of the length. On the balustrade to our left, throned high above the dinosaurs and other treasures of the courtyard, is a shiny brass globe slightly bigger than a basketball, with a little label underneath. Reading the label we learn that this is part of an accurately scaled model of the Sun, Earth, and

Moon with the appropriate distances between them, and that the rest of the model is on the balustrade on the opposite side of the court. Of course, as the court measures some 33 metres across, we can't see any of it from where we are, so we obediently retrace our steps to the other side and find there a little showcase which we would certainly have overlooked had it not been for the brass ball and its label.

Well, of course you know that our planet is a lot smaller than the Sun, and quite far away from it, but both these concepts are difficult to imagine on their own, and even more so in combination. The drawings in books tend to either show the right sizes, or the right distances between the Sun and its planets, never both. So I am warning you that you will probably be shocked when you look into the box on the other side, only to find a pinhead of less than three millimetres diameter representing Earth, with an even smaller pinhead (less than one millimetre) for the Moon just eight centimetres away, on the backdrop of an irritatingly featureless black cardboard square. Looking back across the gallery you still see the brass Sun, but the space between the two balustrades suddenly appears vast and empty.

On the scale of this model, the neighbouring stars might be in some other museum in New York or Tokyo. Just imagine the task of seeing from Oxford whether they have habitable pinheads in Tokyo! Of the other planets in our system, Mars would scrape round the corners of the museum building, while Pluto would be some 1.3 kilometres away. Incidentally, the average distance between Neptune and the Sun misses my house by 30 metres. Taking the elliptical shape of Neptune's orbit into account, I could probably build an extension of the model by installing an 11 millimetre marble in my front porch for casual visitors.

So, right now, we are sitting on this pinhead and we ask ourselves what the shiny ball on the other side is made of, why it came into being, why it shines, and why there are these little specks circling around it which we call planets and consider to be important, although between them they make up less than 1 per cent of the mass of the Solar System. We shall see that we are incredibly lucky to have our Earth speck in the place where it is.

Goldilocks and the three planets

Our blue planet Earth is an oasis of life in the cold immensity of the Universe. Just remember the pinhead in the museum gallery, and then think that the nearest other star, and hence the nearest place with even a small chance of harbouring diverse life, would on this scale be a football somewhere near New York. We're on the only habitable pinhead in a sphere of at least the 1.5-fold diameter of the Earth. Even if there are many other Earth-like planets outside this sphere, life will still be a very small exception in a largely empty and lifeless Universe. This remarkable circumstance needs an explanation. Why Earth, rather than Mars, Venus, or Europa?

It is true that, in comparison with all other celestial bodies we know, Earth appears as just the perfect place. The wide variety of chemical elements, the abundant supply of liquid water, the atmosphere, the protective ozone layer, the slowly changing continents, all the major features of our planet appear to be ideally suited for the task of developing and sustaining a rich and diverse community of living species like the one we observe today.

However, this should not fool us into thinking that Earth has always been a greenhouse, just waiting for seeds to be placed in it so it could be filled with life. Conditions have not always been as gentle as they are today, and when the planets formed it was not obvious that this one was special. Like its immediate neighbours, Mars and Venus, early Earth was an inferno of volcanoes and was battered by meteorites. When the crust had solidified and the meteorite bombardment ceased, adolescent Earth was probably still quite similar to the neighbours. All three may have come up with primitive life at that time. But while life on Earth took over the planet and drastically changed the composition of its oceans, sediments, and atmosphere, our two neighbours went different ways. Mars lost its atmosphere and therefore became too cold to sustain a substantial biosphere, and Venus had too much of a good thing, with the greenhouse atmosphere leading to unbearably high temperatures.

There are many reasons why things turned out 'just right' for Earth, but not for any other planet in the Solar System. Apart from the chemical composition and the size of our planet, the distance from the Sun is one of the most important criteria. The amount of energy received by a given surface area exposed to the Sun declines with the square of its distance from the Sun (because the light is spread out over the surface of a sphere, which is proportional to the square of its radius). Thus, a planet twice as far away from the Sun as we are (i.e. in the asteroid belt beyond Mars) would only receive a quarter of the light and warmth that we get, while a planet orbiting at half the distance would get four times as much. We are located within a relatively narrow band of possible paths around a star, at a distance where the energy supply from the star allows for the presence of liquid water at the planet's surface and a biosphere like ours, which is called the Goldilocks zone (after the Victorian fairytale 'Goldilocks and the three bears'), because it is 'just right'.

When astronomers search for planets in other solar systems (which can only be detected by indirect means so far), they are most interested in objects that are likely to be within the Goldilocks zone of their stars. As the gravitational pull a planet effects on its star is so far the only clue that we can get about the existence of other planetary systems, most of the few dozen planets discovered since 1995 are Jupiter-sized giants whizzing round their stars on a Mercury-style orbit in a matter of days. This doesn't exclude the possibility that each of these systems could have additional Goldilocks planets, but so far we just cannot tell.

Being at the right distance from a suitable star is what can make a habitable planet a living planet. Even though the Jovian moons Europa, Ganymede, and Callisto—located well beyond the Sun's Goldilocks zone—are believed to possess vast oceans that might harbour primitive life, these are permanently covered by thick ice, which practically rules out the evolution of a complex and diverse biosphere like ours. It appears that those cultures that identified the Sun with a life-spending deity did have a valid point (see Chapter 6). We are very lucky in that we are in the right place, and that the time is right too. The early Earth was constantly battered by meteorite impacts, while the late Earth, in a few billion years from now, will lose its atmosphere and water as a consequence of the imminent death of our star. The Jovian moons may become more hospitable than Earth at that point. In about five billion years from now the Sun will begin to run out of fuel, which means that it will first expand into a red giant, roasting our planet to death, and then the fire will eventually go out and the Solar System will cool down to the temperature of the surrounding Universe, just a couple of degrees above absolute zero. In between, there is a window of opportunity of about eight billion years in which a diverse biosphere can flourish on our planet, fuelled by the gentle light of our central star.

Let there be light

The light we receive from the Sun is around eight minutes old, that from neighbouring stars up to a hundred years. Light from other galaxies may travel for millions of years before it reaches our eyes, offering us a window into the past of the Universe. But there are some light particles (photons) that are even older than that and give us insight into the state of the Universe just after the Big Bang, at the time when high-energy particles combined to form atoms for the first time. This kind of light shines at much longer wavelengths (lower energy) than the kind that we can see. It is in the microwave part of the electromagnetic spectrum. As it is uniformly distributed in all directions of the space around us, and as astronomers look at it for revelations about the history of the cosmos, it is known as the cosmic microwave background, or CMB.

For the first few hundreds of thousands of years of its existence the Universe was a very boring place for everyone except particle physicists. It was so hot and dense that the atomic nuclei of hydrogen and helium, which had formed in the famous 'first three minutes', could not combine with electrons to form neutral atoms, but they constantly bounced off each other in a so-called plasma. Apart from highly energetic photons from the gamma-ray part of the spectrum, which were included in this high-energy soup and were constantly swallowed up and spat out by the particles they encountered, there wasn't anything that we would recognize as light.

The Universe came to its senses around 300 000 years after its

Helium, the Sun element

I N THE MID-NINETEENTH CENTURY, the existence of atoms was an unproven hypothesis. John Dalton (see FLASHBACK on p. 120) and the other founding fathers of modern chemistry had shown that the assumption that atoms combined in fixed proportions to form molecules would neatly explain the proportions observed in chemical reactions in the gas phase. Thus, one volume of oxygen gas reacted with two volumes of hydrogen to form two volumes of water vapour. It was a bold step from there to the conclusion that each of these equivalent volumes corresponds to a constant number of molecules, and that the proportions correspond to the numbers of atoms per molecule. For the time being, direct evidence of these atoms and molecules was non-existent.

But then, in 1859, the German physicists Gustav Kirchhoff and Robert Bunsen developed an instrument that allowed them to record characteristic fingerprints of atoms. In principle it was just a Newton-style prism with a camera. If you make a metal wire glow and guide the light through the prism and onto the film, you don't get a continuum of colours as Newton did. Instead, you get a distinct, well-defined array of light stripes, like a supermarket barcode. This so-called line spectrum of a light source is in fact a fingerprint of the chemical elements contained in the source. It tells us which kinds of atoms make up the glowing wire.

And, more interesting than glowing wires, the technique allows you to analyse the composition of light sources that are very far away, like distant stars or like our very own star, the Sun. The lines are so sharp and precisely localized that they allow individual elements to be identified from mixtures containing up to dozens of different kinds. What's more, if you observe lines that don't fit any known element, you know you've found a new one. In 1868 Norman Lockyer and Edward Frankland took advantage of a solar eclipse to analyse a line spectrum of the light coming from the corona. Apart from the well-known and very strong signals assigned to hydrogen atoms, they also found an unknown element, which Lockyer called helium after the Greek name for the Sun, *helios*.

More than 20 years went by before the Sun element was also discovered closer to home, down on Earth. In 1890, scientists analysing uranium ores by dissolving them in acid detected a gas which Sir William Ramsay identified in 1895 as a member of the newly discovered group of the noble gases. (Argon had just been discovered a year

before, and the others were found in the following years, after Carl von Linde succeeded in cooling air to the liquid state and separating its components on the basis of their different boiling points.) Like the Sun helium, the gas found in uranium ore is the product of a nuclear reaction, but this time the reaction is radioactive decay rather than fusion.

Although helium is the second most abundant element in the Universe (hydrogen is the first), it is among the rarer elements on Earth. This is simply due to its low weight. As it doesn't form molecules and its atoms are a lot lighter than all possible molecular species except hydrogen molecules, it is most likely to rise to the upper layers of the atmosphere, when it's free to do so, and eventually escape into space.

If you've ever let go of a helium-filled balloon, you have contributed to our planet's continuous loss of this rare element. What little we have of it is in fact only passing by. The amounts produced by radioactive decay are balanced by the amounts lost to outer space.

While helium is naturally found in places that are either very hot or contaminated by radioactivity, it has ironically found its most important applications in the technologies involved with very low temperatures. Helium is the only element that cannot be frozen at atmospheric pressure, and it only liquefies at four degrees above absolute zero (4 kelvin). Extremely powerful electrical magnets, such as those needed for magnetic resonance imaging and for nuclear magnetic resonance spectroscopy commonly contain superconducting coils cooled by liquid helium. If you cool it even further, at 2 kelvin (K) it converts to a very strange state called a superfluid. It will then flow through tiny capillaries with no friction and conduct heat better than any metal.

beginning, when it had expanded and cooled down sufficiently to let light and matter go their separate ways. As the photons lost part of their energy and moved towards the visible part of the spectrum, they were no longer hot enough to rip atoms apart, so nuclei and electrons could combine to form the two lightest kinds of atoms, hydrogen and helium, which are still the dominant types of matter in the Universe at large. The photons which were released at that time and are still roaming around are what we now observe as the CMB. Over billions of years they gradually cooled off, to the very low-energy range of the microwave bands they now populate.

Observing such an evenly spread and low-energy radiation is not easy. Its existence was first predicted in 1948 by the Russian-born physicist George Gamow (1904–68), one of the pioneers of the Big Bang theory and the author of popular science books still in print today. Nevertheless, CMB was only discovered 17 years later, when Arno Penzias and Robert Wilson at Bell Laboratories spent some time trying to de-bug a microwave antenna designed to monitor satellite transmissions: at first they were sure that the excessive background they were picking up must arise for some humdrum reason, such as the pigeons nesting in the antenna horn! After carefully eliminating all possible terrestrial sources of this 'hum', they finally concluded that it was arising in space, and with the help of researchers at Princeton, who had set out to find this radiation, realized that it represented the cosmic background radiation predicted by Gamow two decades earlier. (Penzias and Wilson were each awarded a third of the 1978 Nobel prize in physics for this 'noise'.) Now, the most interesting feature of the CMB is its total lack of interesting features. If you scan it across the entire sky, the biggest differences in energy you will observe amount to less than one in a thousand, suggesting that the Universe at the time of decoupling was essentially featureless. Furthermore, the way in which the photons are spread out over the wavelength range they populate obeys the predictions of the physical laws about radiation more perfectly than any other phenomenon. For the physicist, a 'black body' is an idealized object that does not reflect any radiation, but only radiates because of the temperature it has. Just as a glowing metal emits visible light (note that the colour of the glow can be used to estimate the temperature), colder bodies send out invisible photons of lesser energy and longer wavelengths. For a given temperature, one can calculate the spectrum (wavelength distribution) of the 'black body radiation'. CMB fits the calculations for a perfect black body kept at 2.725 K. One of the reasons that the interpretation of the CMB as the 'echo of the Big Bang' is accepted beyond reasonable doubt lies in the fact that nobody could come up with an alternative explanation for the existence of a radiation so evenly distributed and so perfectly regular.

Attempts to catch what little variation there may be have relied on both ground- and space-based instruments. Significant progress was made in 1992 when the COBE satellite (COsmic Background Explorer) scanned the entire sky at a resolution of seven angular degrees. It has to be said, however, that a similar scan at visual wavelength would miss the outlines of the Sun and the Moon (apart from noticing a vague brightness in the sky), as their apparent diameter is only about half a degree. Thus, valuable as the COBE result was, it left researchers yearning for more data at higher resolution.

Ground-based observatories have achieved higher resolution for the very limited segments of the sky they have had access to. Half-way up, a probe based in a balloon circling the Antarctic (BOOMERANG for Balloon Observations of Millimetric Extragalactic Radiation and Geophysics—I hope this acronym falls back on whoever invented it!) has monitored its part of the sky at 0.25 deg. But with the successful launch of the NASA probe MAP (Microwave Anisotropy Probe) on 30 June 2001 (designed to achieve a resolution of 0.3 deg.), and the scheduled launch of the ESA observatory Planck in spring 2007 (0.17 deg.), each of which will cover the entire sky, space-based observation of the CMB will enter a new era and hopefully reveal what little structure there was in the nascent Universe. Cosmologists also hope that the analysis of the fine structure of the CMB will allow them to answer some fundamental questions about the space-time fabric of the Universe.

The Universe cannot have been entirely featureless some 300 000 years after time zero, because then we wouldn't be here to talk about it. The inhomogeneities which researchers are hoping to map by studying CMB with their improved instruments must have something to do with the fact that the Universe came up with lumps that condensed into stars and galaxies. Shame that nobody was there to watch when the first ever star lit up, and visible photons started their travel to infinity. To those first stars we owe all the elements heavier than helium (as will be explained below). Without them, i.e. before the first supernova explosions, there would have been no possibility of life anywhere. We are all, as some writers have poetically pointed out, the children of those long-extinguished stars.

Some five billion years ago, somewhere in a galaxy known as the Milky Way, a loose cloud of gas particles, possibly the scattered leftovers of a supernova explosion, collapsed into a ball with a thin flat rotating disc around it. The ball contained enough mass for the gravitational energy to fire up the atomic reactions which make stars shine, and thus a quite normal yellow star was born—the one we call the Sun. During the first few million years of its new life as a star, the mass compacted further, and the star shone several times brighter than in the very beginning. The mass contained in the rotating disc also collapsed into lumps which eventually became the planets we know today. For the overall bookkeeping of the mass contained in the Solar System, however, the planets are mere peanuts: 99.9 per cent of the mass is contained in the central star.

Since then, the disc surrounding the young star has consolidated somewhat, forming (at least) nine planets with their 30-odd moons. Most of the stray bodies that made life in the Solar System dangerous in the first billion years or so have either hit a planet, found a safe orbit, or left the Solar System. The energy output of the Sun has risen by another 25 per cent, but life on Earth has found ways of maintaining its overall environment in a life-enhancing shape. Although catastrophic impacts can still happen, and although the celestial mechanics is not quite as predictable as Newton would have thought, our planet has a fair chance of sustaining diverse life for another four billion years, as long as we humans don't do anything too silly. Provided that life and some kind of civilization do survive through to the end of the Sun's fuel reserves, an interstellar ark will then be required to transfer our biosphere to a younger star.

But let us leave this problem for science fiction authors and have a look at the processes which keep the solar fire burning.

The fire within

Atoms, those allegedly indivisible (in Greek: *a-tomos*) fundamental building blocks of matter, soon after the discovery of their physical reality proved to be a lot more divisible than their name suggests. Firstly, they are made of negatively charged electrons and a positively charged nucleus. Electrons can be removed from most atoms quite easily—that is what chemical reactions are all about. Then, atomic nuclei have two kinds of building blocks: the positively charged protons and the uncharged neutrons. Certain kinds of nuclei, especially those which are quite heavy and rich in neutrons, have been shown to be unstable—they decay by forming lighter atoms and emitting radioactivity. In 1938 the German physicists Otto Hahn, Lise Meitner, and Fritz Straßmann found that heavy atoms can be split by bombarding them with nuclear particles.

If you list atomic nuclei by increasing weight, the most stable ones will be in the middle, around iron and nickel. For this reason, energy can be freed by splitting atoms much heavier than iron, but not, for instance, by splitting light atoms such as aluminium. However, energy can be produced by merging some of the lightest atoms. While the nuclear fission process can be technically mastered (at least in some places, and most of the time), the latter, known as nuclear fusion, has been used in hydrogen bombs, but not yet tamed in a way to provide energy in a controlled reaction and in a process that is economical overall. And still this is exactly the kind of energy we are getting from the Sun. Solar energy is exclusively produced by nuclear fusion.

Hydrogen is the most abundant element in the Universe, and it is also the lightest element, as far away as possible from the very stable iron nucleus. Its neighbour in the periodic table, helium, has an unusually high energy of binding keeping its nucleus together. Therefore, the merger of two hydrogen atoms to form helium can liberate an enormous amount of energy, as it does in the inner core of the Sun. The catch is that the atomic nuclei which are to be merged are positively charged, and hence repel each other like all bodies of equal charge would. Therefore, energy has to be invested into overcoming this repulsion at the beginning, before the much larger energy yield of the fusion reaction can be harvested. This is the reason why nuclear fusion is not yet being used for energy production here on Earth, and why it works so nicely in the Sun and in other stars at extremely high temperatures (up to 16 million degrees C).

The fusion reactor which provides us with heat and light is in the core of our Sun, a ball within the ball. On the scale of the museum model, it would be a little bit smaller than a tennis ball. Because the nuclear reaction is very rich in energy and occurs at very high temperatures, the energy liberated by it does not take the shape of heat and light as in an ordinary fire or light bulb. It is mainly radiated off as extremely short-waved electromagnetic radiation, known as gamma rays. (In this respect, fusion resembles nuclear fission.) A further product of the fusion reaction is a stream of strange elementary particles, the neutrinos, which have very little mass (or perhaps no mass at all: this is still controversial—see FLASHBACK).

While the neutrinos, which are hardly ever thrown off their track no matter what obstacle arises, escape through the outer shells of the Sun within seconds and just keep going straight ahead, the gamma rays on their path will interact with countless hydrogen and helium atoms, thereby losing their direction and part of their energy. Thus the energy emitted as gamma radiation in the core of the Sun needs thousands of years to penetrate to the surface, from where it is sent out into space as a much less energetic mix of visible light, heat, and ultraviolet light. The layer which sends out this radiation, which is essentially our sunlight (except for those parts which our atmosphere filters out), is called the photosphere. It is much colder than the core, at only around 6000°C, and reaches from a depth of 1.8 to 1.9 millimetres in the model (see legend to Figure 1.1 for the 'real' numbers). Between photosphere and core there is also a zone where energy is mainly transported by convection, that is by hot gases rising and cold gases sinking, as in our atmosphere or in a pot of water heated from below.

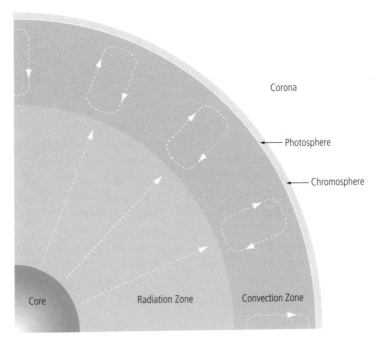

Corona

Photosphere

Chromosphere

Core   Radiation Zone   Convection Zone

● Earth (for size comparison)

Figure 1.1 Cross-section of the Sun according to current models. Distances from the centre to the outer limits of the inner zones are as follows (in kilometres): core, 150 000; radiation zone, 500 000; convection zone, 690 000. The widths of the photosphere and chromosphere are estimated to be 500 and 8000 kilometres, respectively.

The photosphere is in turn surrounded by a thin gas layer known as the chromosphere (the outer 1.8 millimetres of the brass ball). This layer produces the turbulences of the solar surface which we can sometimes observe. While the photosphere and the chromosphere determine our visual image of the Sun, they are by no means the end of the story. The visible ball is surrounded by an invisible wrapping of gas, the corona, which could be described as the Sun's atmosphere. It expands all the time and in all directions (away from the Sun), a phenomenon which is also known as the 'solar wind'. This wind causes the tails of comets always to point away from the Sun. The corona becomes thinner as the gas particles available

How to catch a neutrino

WOLFGANG PAULI (1900–58) was known as a genius in theoretical physics, and dreaded as a critic who could devastate a colleague with just a few words. In 1930, he found himself deeply immersed in a dilemma without a solution that would have withstood his own critical mind. A nuclear reaction in which a neutron splits up to form a proton and an electron appeared to violate the physical law concerning the overall conservation of energy. Just to save the day Pauli very reluctantly postulated (physicists' lingo for 'invented') a new kind of particle that would be able to carry away the excess energy, but wouldn't contribute measurably to the charge or to the mass balance. This hypothetical particle, which his Italian colleague Enrico Fermi baptized by the Italianate name of neutrino, would engage only in one of the three fundamental forces of physics (the weak interaction). This meant that neutrinos would very rarely interact with anything else, and thus would be extremely difficult to detect.

A quarter of a century went by before these particles could first be detected in the proximity of a nuclear reactor. Theoreticians then put forward another hard-to-prove hypothesis: the Sun, being an immense nuclear reactor, should also send out unbelievable numbers of neutrinos. Calculations carried out in the 1960s suggested that 2 per cent of the energy created by the Sun is emitted as neutrinos. This means that the area of a thumbnail is passed by 100 billion solar neutrinos every second, but without deflecting a single one of them from its path. As planets are no significant obstacle to the progress of neutrinos, they even shoot through your body while you're on the night side of the Earth. But how to detect such evasive particles?

All first-generation neutrino detectors relied on the same principle. A huge amount of a highly pure substance was placed in the neutrinos' path in the hope that they would, by a rare nuclear reaction, turn some atoms of that substance into different, radioactive atoms that would be easy to detect. Thus the Gallex experiment had 30 tons of gallium in a salt brine. Any neutrino reacting with a gallium atom would produce a radioactive germanium atom. In order to be sure that the radioactivity measured came from this reaction, it was necessary to filter out the cosmic background radiation very efficiently. A one-mile layer of granite would do the trick, so the Gallex experiment was placed in a tunnel in the Alps, while other, similar experiments are carried out either in mines or underneath the Antarctic ice shield.

To their frustration, researchers only caught about one neutrino per week in their gallium trap. They would probably have given up had it not been for one day in 1987, when a spectacular 19 neutrinos were observed in a short time. This suggested that the usual 100 billion solar neutrinos per thumbnail and per second were being swamped by a thousandfold bigger number from an unknown source. Hours later, astronomers observed the light of one of the closest supernova explosions to be recorded in the twentieth century. The neutrino shower had been the first messengers of this event. In those hours neutrino astronomy was born. New and improved detectors were developed in the hope of tracking down events in the dark depths of space that remain hidden to optical astronomy.

The most powerful detector of the new generation is called Superkamiokande and contains some 50 000 tons of ultrapure water in a cylindrical barrel surrounded by some 11 000 light detectors. It works on the principle that in the rare event that a neutrino hits an atomic nucleus it can create other high-energy particles, which send out a characteristic flash of light known as Tcherenkov radiation. This comes from particles moving faster than light would move in water (but still slower than c, the velocity of light in a vacuum), and is thus comparable to the bang we hear when an aircraft crosses the sound barrier.

Results from Superkamiokande and other modern neutrino detectors have tended to show smaller numbers of neutrinos, and different ratios between the three different kinds, compared with what theorists predicted. One of the possible explanations suggested is that the kinds of neutrinos can interconvert en route. Only two of the three can be detected by their Tcherenkov light, so the missing neutrinos may have turned into the third kind. In June 2001 researchers from the Sudbury Neutrino Observatory in Canada announced that comparison of their data with that of other observatories revealed that neutrinos do indeed 'switch flavour'. This finding suggests that they have a mass after all, although it would still be many orders of magnitude smaller than that of the electron.

spread out over a bigger sphere, but it is hard to define an outer limit for it. Probably the solar wind ends only when it collides with the corresponding winds of other stars.

Now that we have seen how the Sun works, we can come back to the fate it faces in the future. The solar fusion reactor has still got enough hydrogen atoms to keep burning for a few billion years longer, although it will be getting a serious problem of waste disposal. The helium atoms formed as a product of the fusion reaction are heavier than hydrogen and hence they are accumulating close to the centre of gravity, in the middle of the core. There is a 'nuclear waste dump' at the centre of our Sun, which is growing steadily and is denser than the surrounding hydrogen. Because of its higher density and bigger gravitational pull, this helium core within the core will collapse further. When it reaches a critical mass and density, helium itself will become nuclear fuel, with carbon atoms as the waste product (which will again form a core within the core). The two-step fusion reactor (hydrogen to helium to carbon) will on the outside be a red giant, putting our Earth in a rather uncomfortable position.

When enough carbon has accumulated to fuel higher fusion reactions (leading to oxygen and neon), the end is nigh. Hydrogen is running out, and the nuclear fusion of the heavier elements does not liberate enough energy to counteract the gravitational forces which make the red giant collapse into a white dwarf. (Our Sun is simply not big enough to drum up a supernova explosion, so the elements formed inside it are unlikely ever to benefit any starchildren.) The dwarf casts off its gas layers and keeps glowing at a lower intensity for a few billion years. When the fusion reactions eventually die down, the Sun will turn into a black dwarf and gradually cool down to the temperature of the surrounding empty space.

The observer that was lost and found

All we know about the Sun has been deduced from the detailed analysis of the radiation it sends out. Much of it is outside the visible spectrum, in the high-energy wavelength bands of the ultraviolet and beyond, which our atmosphere thankfully filters out. If solar physicists want to get the full picture, they need to place their instruments in space. And that's exactly what they have been doing over the last few decades. The most recent and advanced satellite to be dedicated to studies of the Sun is the SOlar and Heliospheric Observatory, SOHO, which the American and European Space agencies, NASA and ESA, launched in December 1995.

On 14 February 1996, SOHO arrived at its destination, the privileged point on the Sun–Earth axis where the gravitational forces of both bodies are exactly balanced. Due to the inequality of the masses involved, this so-called inner Lagrangian point is only 30 centimetres away from our pinhead (about 1 per cent of the total distance to the Sun). But then, unlike any satellite orbiting the Earth, SOHO sees the Sun 24 hours per day, and remains completely unaffected by any terrestrial problems (other than controllers' errors). Of its 12 onboard instruments, 6 are designed to probe the Sun's atmosphere, 3 look at its interior, and 3 look at the solar wind. Its location also allows it to serve as an early warning system to report magnetic storms coming our way.

SOHO started its mission at the lowest point in the magnetic activity of the Sun, which changes in an 11-year cycle that is thus far poorly understood. The spacecraft carries enough fuel to keep on watching the Sun throughout one complete cycle, including the 2000/2001 solar maximum and the decrease of the activity towards the following minimum.

The hopes for such a long-term observation were very nearly shattered on 24 June 1998, when a routine maintenance operation left one of the gyroscopes (spinning wheels meant to stabilize the orientation of the craft in space) on the wrong setting. The onboard computer wrongly assumed the probe was spinning around and fired the thrusters to counteract the movement. Control staff responded swiftly by switching off the gyroscope, except that they picked the one that was still working, leaving the probe tumbling out of control. With its solar panels turned away from the Sun, its energy supply was off. A few minutes later SOHO was frozen as if in a coma, unable to move or respond to commands from Earth.

In a painstaking search, operators were able to localize the probe using radiowaves reflected from its solar panels, and to determine its orientation and slow rotation. They worked out that there would be a window of opportunity for reviving the probe two months later, when the solar panels would again face the Sun. Sending very carefully thought-out commands at precisely the right time, they managed gradually to thaw the fuel lines, re-establish communications with the probe, and wake up the scientific instruments. Considering the harsh conditions of spending two months in space without heating, scientists were surprised to find that all the scientific instruments came back to life. The gyroscopes, however, were lost for good soon after regaining contact. The operators made do without them, instructing the probe to orient itself towards the Sun instead. Since then, SOHO has carried on with its mission without major problems, completing the observation of a solar half-cycle (from the minimum to the maximum of activity) in June 2001.

As data from the probe are still coming in and being analysed by researchers, it is not yet possible to give a final summary. Many important findings have already been reported. Seismographic measurements made by SOHO have, for instance, enabled researchers to 'see' sunspots on the far side of the Sun. As the Sun turns round once every 27 days, this knowledge allows scientists to predict a few weeks in advance the magnetic storms that are coming our way once the sunspots have arrived on our side. This is important, because such events can seriously affect satellites, spacecraft, and telecommunication systems.

Research with SOHO's SUMER (Solar Ultraviolet Measurements of Emitted Radiation) instrument has addressed the origins of the solar wind, the high-speed particle stream that constantly blows across the entire Solar System. Previous work had suggested that it comes from patches in the magnetic field around the Sun known as 'coronal holes'. Researchers working with the SOHO instruments studied the flow of ions of the noble gas neon and found that the solar wind appears to flow from the inner rims of these 'holes', at the base of the corona.

Other research areas addressed by SOHO investigators include the question of why the corona is so much hotter than the photosphere underneath it, and what exactly is contained in the core. On

Anders Jonas Ångström (1814–74)

L IKE THE SCIENTISTS MENTIONED in the FLASHBACK on helium, the Swedish physicist Anders Jonas Ångström used spectral analysis to study the light from various sources. In particular he applied spectroscopy to the Sun, the northern lights (aurora borealis), and to electrical sparks.

Anders Jonas Ångström studied at the University of Uppsala, where he also became a lecturer in 1839 and chairman of the physics department in 1858. Using spectral analysis of sunlight, he discovered that hydrogen is present in the Sun. He also compared the spectrum of the northern lights to that of sunlight, and concluded that the aurora is not composed of reflected sunlight. (It comes, in fact, from atoms of our very own atmosphere, which are excited by particles coming from the Sun, especially in years of high sunspot activity.) He also published a detailed map of the solar spectrum in 1868.

All these achievements would not have sufficed to rescue Ångström from obscurity. What did it for him was the bright idea of inventing a unit to conveniently describe all those wavelengths he was dealing with in his work on spectroscopy. He chose it to be the 10 billionth part of a metre (10^{-10} metres, or 0.1 nanometres as we would now say). That was a very lucky choice, as scientists realized around the end of the nineteenth century that atoms have diameters in this range. In 1905, this unit was named in his honour (the angstrom, with no accents, short: Å), and it was used throughout the twentieth century by scientists in disciplines ranging from X-ray crystallography through to astronomy. Ironically, it was quickly replaced by the modern SI unit (1 nanometre = 10 Å) in Ångström's own field, spectroscopy, where the wide range of wavelengths to be studied, sweeping many orders of magnitudes, is most easily covered by a strictly systematic set of units such as the metric one. His fame lives on because researchers dealing with atoms, e.g. X-ray crystallographers specifying the resolution of a molecular structure, still use the Å.

top of that, SOHO has also become the world's highest-scoring comet finder, with the discovery of more than 300 new comets by February 2001. Many of these, however, are now no longer with us: they were on a collision course with the Sun.

A planet in the balance

But let's leave SOHO keeping its watch and return to Earth, to see how the Sun enables our planet to grow living things. Now we have got a planet circling a suitable star at the right distance and receiving the right energy doses. But still, these conditions, though necessary, are not sufficient to guarantee a climate in which the life-enhancing effects of light can unfold.

If, for instance, the wrapping artist Christo were to become bored with packaging historic buildings such as the Berlin Reichstag, and endeavoured to wrap our whole planet in white plastic foil, most of the solar energy coming our way would be reflected back directly into space. The small part that could penetrate or be absorbed by the wrap would not be sufficient to stop the oceans from freezing. Thus liquid water—a major requirement for life—would cease to exist. In a more likely scenario without the need of an artist, a small increase in the extent of the polar ice caps could increase the fraction of the sunlight reflected directly into space gradually (scientists call this fraction the 'albedo' or whiteness of the planet). The reduced energy intake would cool our climate down, which would lead to a further growth of the ice caps, and so on. According to a controversial theory proposed by Harvard geologist Paul Hoffman in 1998, a runaway glaciation period may have actually taken place between 750 and 550 million years ago. Volcanic emissions of the greenhouse gas carbon dioxide would have eventually saved 'Snowball Earth' from remaining frozen for ever.

If, in contrast, the artist set his mind on wrapping the planet in clear plastic, which lets the light in but stops excess heat from getting out, most of the solar energy would be trapped in a global greenhouse. Ice caps would melt and thereby make things worse. Eventually the oceans would evaporate and Earth would become a lifeless hothouse similar to Venus. Again, such a catastrophe could conceivably happen even without an artist, triggered by green-house gases or a reduction of the polar ice caps for whatever reason. The large amounts of methane trapped at the bottom of the oceans in the form of methane hydrate ('burning ice') would be more than sufficient to trigger such a change.

These simple thought experiments show how sensitive the balance of our global climate is. Living beings will both contribute to this balance (on either side) and depend on its stability for their survival. Plants, for instance, remove the greenhouse gas carbon dioxide from the atmosphere and produce oxygen instead, which animals like ourselves need for breathing. We, on the other hand, eat plants and use the oxygen from the air to burn the carbon they have accumulated to form carbon dioxide. If the plants worked a lot harder than the animals, the oxygen content of the atmosphere might increase, leading us to a situation where ordinary materials such as paper or leaves would become dangerously flammable. If, on the other hand, the animals ate up the plants faster than they could grow, there might be a shortage of oxygen and a greenhouse effect from excess carbon dioxide.

It is intriguing that in spite of the possibilities for catastrophic runaway changes the composition of our atmosphere and our climate has (probably) remained fairly constant for half a billion years. The ice ages are in fact only a faint echo of what might have happened (and what quite possibly happened some 700 million years ago). The British chemist James Lovelock realized how much the physical conditions prevailing on our planet are influenced in an obviously positive, life-enhancing way by the activities of the lifeforms inhabiting it. He cast this observation into the hypothesis which is now known as the Gaia theory, suggesting that the whole planet, including the biosphere, acts like a self-regulating system or living being, which he named after the Greek Earth goddess, Gaia.

In one thought experiment which he used to illustrate his theory, Lovelock populated a model planet with two species of daisy, a white one and a black one, with the former thriving better at higher temperatures, the latter at lower temperatures. When the climate warms up, white daisies spread more widely, leading to an increased albedo and reversal of the warming trend. Conversely, when it gets colder, black daisies take over and help the planet to collect more solar power. The whole process is a negative feedback loop, which essentially ensures that any deviation from the 'normal state' of the system triggers a response which counteracts the deviation. Similar, if vastly more complex, self-regulation mechanisms may be the true explanation for the surprising fact that the Earth's climate has stayed in the optimal range in spite of changes in the solar energy output.

In these regulatory cycles, the negative feedback loops which stabilize the status quo must be stronger than any positive feedback loops (like the albedo/ice-cap effect) which would favour catastrophic changes. What is happening on a global scale appears to be similar to the way in which living organisms keep their internal conditions (body temperature, pH, water content) approximately constant, which is why Lovelock identified the whole Earth as a living being. It also resembles technical regulation processes or, for instance, the way you would steer a car. If you are a safe driver, you constantly counteract small deviations from the straight intended course. In fact, every vehicle proceeds in more or less sinewave-shaped curves rather than straight lines, as small deviations from the intended direction have to be corrected for (as will be discussed in more detail in Chapter 4). In this comparison, the Earth could be a self-regulating cybernetic system, after the ancient Greek word for a helmsman.

Without getting too involved in the Gaia discussion, let us just have a quick look at the energy balance of our planet. All the energy available to living beings is ultimately derived from three sources: Earth, Moon, and Sun. The smallest factor among these is the input of the tidal forces. Gravitational pull and centrifugal force are exactly balanced only at the Earth's centre of gravity and in those places which are at the same distance from the Moon. If you hold a wire in both hands and use it to cut a soft fruit (the Earth) into two equal halves (one facing your head—the Moon—the other pointing away from it), the cut will show you where the two forces are in equilibrium. In other words, the balance is right wherever you see the Moon just about rising or setting. On the hemisphere closer to the Moon, the oceans are pulled towards it a bit more strongly, while those on the opposite (moonless) side experience a slight excess of centrifugal force over the Moon's gravitational pull. Therefore the water is a few metres higher at these two extremes and lower in the middle. This is why we see high and low tide twice in 25 hours at the coasts. (If, as one could erroneously imagine, all the water was attracted to the Moon's side, one should observe high and low tide only once per day.)

This also means that, twice every day, the coasts are being pushed against these mountains of water, which results in mechanical friction. You may not have noticed this so far, but this friction actually slows down the Earth's rotation by two-thousandths of a second each year, as the motion energy is turned into heat, in much the same way as braking a car results in heating up the brakes and the tyres. As spinning tops go, a planet is quite a big one, able to carry a lot of energy even if it only turns round once a day. Hence even a very minor change in its speed involves a considerable amount of energy—in this case it's 2.7 terawatts (trillion watts or 10^{12}), which corresponds roughly to a quarter of the total energy consumption of humanity (see Figure 1.2 for a graphical illustration

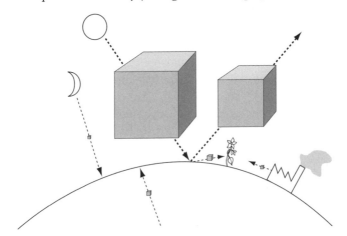

Figure 1.2 Illustration of the energy balance of the Earth. The cube volumes are representative of the amounts of energy flow (in terawatts, TW): tidal energy (from the Moon) 2.7; humanity, 10.8; geological sources, 32; from the Sun, 173 000 (of which 58 000 are reflected); sunlight used by plants, 43.3.

of all the big numbers mentioned in the following text). Not that there is any chance of using a significant part of this energy.

A ten times larger flow of energy is emerging from the interior of our planet. Part of it is residual heat stored in the liquid core, another part results from natural radioactivity. Processes like hot lava flowing out of rifts or volcanoes are heating the surface with 11 terawatts; heat simply conducted through the upper layers provides another 21 terawatts. Taken together, the heat leaking out of the Earth's bowels makes up three times humanity's needs.

If you think these are big numbers, think again. All the sources mentioned so far only provide a ten-thousandth of the energy which the Sun gives us. One hundred and seventy-three thousand terawatts are burning down on us, of which roughly a third is reflected straight back into space. Half of it serves for heating land, water, and air, thus compensating for the heat that we constantly lose to outer space. A little less than a quarter goes into vaporizing water, and thus ultimately into heating the atmosphere, when the water forms raindrops again. Plants only use one four-thousandth of the solar energy available, which is still four times more than what we use. With this energy, plants feed almost the entire biosphere, including, of course, six billion people.

Most of the sunlight goes out into empty space and may travel for billions of years before it hits any object. Only a little more than a billionth of it comes our way and, of this, one four-thousandth is enough to feed the entire biosphere. Just think how many copies of our biosphere the Sun could keep alive! (The British–American physicist Freeman Dyson once suggested that a much more advanced society than ours might wish to harness all that energy by taking Jupiter apart and using the material to build a sphere that encapsulates the Sun and catches all its radiation.)

No life without light?

While this chapter has hopefully made it clear that our planet is a perfect place for life to evolve and thrive in, the question of how it came to be here has remained a mystery. We don't really know whether the first cells drifted around in a sunny little pond or in dark, volcanically heated clefts of the sea floor. It is certain, however, that only the combination of a generous supply of solar energy with an atmosphere which can retain a major part of this energy could guarantee the relatively constant climate that we have enjoyed over most of the past four billion years, and which has provided ideal conditions for life to spread on Earth to the point of dominating and reshaping the whole planet.

Light is the part of the wide range of electromagnetic radiation phenomena that we can see and therefore distinguish from the infrared (heat) rays on the long-wavelength side, and also from the potentially dangerous ultraviolet radiation on the more energy-rich short-wavelength side. One should bear in mind, however, that this definition is entirely artificial and based on the limits of our visual sensory apparatus (to which we shall return in Chapter 6). Cats and bees, if only they could talk, would define light very differently. Three billion years ago there was no basis whatsoever for such a differentiation. Like today, a broad range of electromagnetic waves shone down onto the young Earth from the Sun, and as there were neither eyes to see any light, nor photosynthesis to make use of its energy, this radiation was useful for life only in an unspecific sense, in that it provided heat and—in the ultraviolet range—even energy densities which could trigger chemical reactions.

Photosynthesis, then, was a major technological revolution—the biggest that our planet has ever seen. Apart from providing a direct link between solar energy and chemical energy that can be used in metabolic reactions, it also changed the face of the planet for ever, as we will find out in due course. So big was the change brought about that nearly all lifeforms became addicted to sunlight—either directly or via some product of photosynthesis, such as oxygen. Without light, life would be impossible now.

CHAPTER 2

The oxygen revolution: how cyanobacteria changed the world

JUST AROUND THE CORNER from the University Museum, there is another favourite haunt with some scientific connotations, the University Parks. It has around 30 hectares (but shrinking, as the Science Area grows) of well-groomed lawns, the unavoidable cricket grounds, and a scattering of solitary trees that each spring allow you to watch the spectacular unfolding of their genetic plan, as buds turn into leaves and flowers.

As a boy interested in maths and physical sciences, I didn't think much of plants. I thought of botany as a descriptive science more akin to stamp collecting, a suitable pastime for people like my grandparents, who were into gardening and orchid-spotting. It was only after learning some genetics and biochemistry that I came to admire plants for the way in which they publicly display the miracles of development and differentiation, which mammals like us hide in the womb. Far from being soft, descriptive science, this turned out to be a precise series of events driven by a mathematically well-defined algorithm, creating complexity on a visible scale from a set of invisibly small elements. Somebody should have told me earlier.

The other thing that we should admire plants (and certain bacteria) for is that they keep us alive by harvesting the energy of sunlight and turn it into nutrients that all life depends on. But how did this dependence come into being? At the starting point (described in Chapter 1) we have a light source and a habitable planet orbiting it at the right kind of distance. What we need to get our investigation of 'light and life' going is some life. Turning a previously sterile planet into one inhabited by life is far from easy, be it by spontaneous evolution of chemical processes into bio-chemical ones, or be it by seeding from space. How it actually happened on our own planet is one of the hard problems that scientists are facing today. In the first section of this chapter we shall discuss some of the current theories and what little evidence there is for or against them.

Once life had started to thrive on Earth light and life led rather separate existences, as the first organisms did not have the right tools to make any specific use of the sunlight—except for the unsophisticated use of its energy dispersed as heat. Only a few hundred million years later the first antennae for sunlight evolved, descendants of which are still found in certain bacteria. Later, a second, more efficient way of using sunlight was developed, which had the slight disadvantage of producing a byproduct which was very toxic for all organisms then alive: oxygen. Eventually, how-ever, this was the breakthrough step which allowed the evolution of higher lifeforms and the biosphere we know today. Photosynthesis and its impact on life on Earth are the focus of this chapter.

Let there be life

In the early 1860s, when Louis Pasteur (1822–95) was already the most famous scientist of his era and had all the world talking of those microbes he found in wine, beer, milk, and just about every-where, he faced the toughest problem of his career: where do the microbes come from? Pasteur believed that they were floating around like dust, and could drop into anything exposed to the air. Many of his contemporaries, however, believed that, on a suitable nutrient, bacteria could pop into life from nothingness—a hypo-thesis known as the 'theory of spontaneous generation'. Pasteur then carried out the crucial experiment to disprove this notion: he sterilized medium in a sealed vial and showed that it remained sterile for weeks and months—no trace of bacteria arising sponta-neously. He also sterilized air and mixed it with the sterile medium, again obtaining no evidence of microbial life.

Not quite satisfied with this evidence, his opponents demanded that he should allow the medium to react with untreated air, believing that microbes could originate from a chemical reaction between the two. Pasteur's former boss and mentor, the chemist Antoine-Jérôme Balard (1802–76, the discoverer of bromine), gave him the crucial idea. A vial with a long neck, winding up and down like that of a swan, would allow the air to access the medium, while any bacteria suspended in it would drop out in the neck region without reaching the liquid. The demonstration succeeded and brought final proof (convincing at least the Académie des Sciences, if not the most stubborn among the supporters of spontaneous generation) that microbes don't pop up from nowhere, but that they sail through the air like dust particles and start feeding and growing wherever they drop onto something edible.

But hang on, there must be a rub somewhere. When our planet first formed some four and a half billion years ago it was certainly sterile, but just a couple of hundred million years later there were microbes around which had apparently come from nowhere. So spontaneous generation must have worked at least once. Why this miracle worked once upon a time but cannot repeat itself today, and the exact way in which it worked, are some of the trickiest questions biologists are facing.

Ninety years after Pasteur's experiment with the swan-necked flask, another chemist tried to bring about spontaneous generation in a bottle. This time the aim was not to show it was impossible, but rather to find the right conditions which enabled the process to occur in the very early days of our planet. Stanley Miller, working as a graduate student in the lab of the atmospheric chemist Harold C. Urey, created a miniaturized version of Earth before life in a flask.

His apparatus included a miniaturized 'ocean' and an 'atmosphere'. While our present atmosphere is rich in oxygen and therefore can further processes like combustion or rusting, the primeval atmosphere was the exact chemical opposite. In the mixture of methane, ammonia, hydrogen, and water vapour that surrounded the young planet, any rusty nail would become shiny and any flame would extinguish. (In chemical terminology, the primeval atmosphere was reducing, while ours is oxidizing.) Through the primeval atmosphere in his flask, Miller sent lightning created with the help of electrodes incorporated in the apparatus, and he made the ocean boil for a few days. Each day he extracted a small sample of the liquid and analysed its composition.

Nobody would remember this experiment today if the reddish goo that started to form in Miller's primordial soup within a couple of days had contained only inorganic molecules and simple hydrocarbons. What made Miller and his experiment famous was the discovery that in the boiling flask substantial amounts of more than a dozen different amino acids had formed, including six of those which are the essential building blocks of proteins in all organisms alive today.

All at once, the idea of life originating on the newly formed planet covered by boiling oceans, and shaken by frequent volcanic eruptions and meteor impacts, did not seem that implausible. If important building blocks of life can form under these conditions, it was only a matter of fine-tuning the conditions to allow the construction of giant molecules that can carry information, interactions between them, and a way by which the information can be passed on to future generations. Once a system is in place where molecular information can be inherited with small errors, and natural selection can take its pick, evolution kicks in and explains neatly the remaining three point something billion years of natural history.

Thus it was with great enthusiasm that Miller and others embarked on the endeavour to improve upon this seminal experiment to produce more complex molecules which can carry genetic information, but the task has not yet been completed in a satisfactory manner. Today scientists are probably less convinced that this enigma can actually be resolved, although this does not stop them from making new theories. Without entering too much into the details of the discussions about the various theories and hypotheses that scientists are pondering, I shall only mention two important concepts representing major trends in this research. Firstly, researchers trying to improve upon the primordial soup setup have found that adding minerals to the recipe can enhance the potential of this experiment significantly. Scientists may disagree on the question of which mineral may have been the key to the origin of life, but many will agree that some kind of interface between a solid mineral and a liquid solution is likely to have hosted some very important steps in the transition from simple chemistry to complex biological systems capable of evolution.

The second concept that should be mentioned here is the so-called RNA world scenario. This refers to a hypothetical precursor generation to modern-day cellular life. According to this scenario, the modern division of tasks between proteins (function), DNA (information storage and inheritance), and RNA (information transfer and some function) was preceded by a simpler setup where only one kind of molecule was in charge of both function and information. As RNA is known to be good at both, it would be the best candidate for such a primeval biomolecule. This theory is far from being proven, but even if one accepts it as a plausible explanation, there is still a gap on the time axis between the making of the first simple building blocks by primordial soup chemistry and the first RNA 'organisms'.

There is no certain knowledge about the events which led to the first microbial cells, and how these evolved before they split into the family tree of species surviving to this day. The history books of molecular evolution begin with the last common ancestor of all creatures alive today, a microbe affectionately known as LUCA (Last Universal Cellular Ancestor). LUCA was certainly a quite complex and sophisticated cellular organism, and it was probably several hundreds of millions of years younger than the very first cellular lifeforms, some of which are its ancestors, while others have simply become extinct. LUCA's descendants split up into three major family branches. The one that we belong to, along with all other multicellular beings and some single-cell organisms such as yeast and algae, is known as the domain of the eukaryotes. This label indicates that our cells have a separate compartment to store our DNA, which is known as the nucleus. The second includes common bacteria, such as the well-known *Escherichia coli*, and many pathogens of infectious diseases. The third, less well-known branch is that of the archaebacteria or archaea, which although they may resemble bacteria in shape and size, show important differences in the molecular makeup of their cells. Archaea have been very successful in colonizing extremely hostile environments, where high temperatures and pressures, along with chemical stress factors, make life difficult.

Model organisms from each of the three domains of life have been studied in great detail, including the recent sequencing of dozens of complete genomes. Although it has emerged that genetic material was transferred between these major branches, which makes any analysis based on individual genes highly risky, it is fairly clear that any genetic traits that are found in representatives of all three branches are quite likely to arise directly from the common ancestor. Thus we can be sure that LUCA, like all of today's cellular lifeforms, stored its genetic information as DNA, that it used RNA as an information carrier, and that it had ribosomes, which made proteins following the genetic information conveyed by RNA. Dozens of proteins have been part of LUCA's elementary toolkit. However, there was a limit to what could be achieved with this elementary setup of a bacterial cell. Most importantly, bacteria have never been able to evolve into multicellular organisms. Before higher organisms could enter the scene, a new kind of cell had to be invented.

How to make complex cells

Baker's yeast, goldfish, daisies, your neighbour's dog, you and I, all share a rather important genetic trait: our cells are of the type known as eukaryotic. Such cells are typically 10 times bigger (in each dimension!) than those of bacteria and archaea, so they comprise roughly a thousand times the volume. Unlike the bacterial cell this volume is subdivided into compartments which are enclosed by membranes, and each serves a specific function (Figure 2.1). The compartment which originally served as the defining trait of eukaryotes, namely the 'true nucleus', for instance, encloses the bulk of the organism's DNA. Other compartments are in charge of creating energy from burning nutrients (the cellular 'powerstation' known as the mitochondrion), or of converting sunlight to energy, which is the business of the chloroplasts found in plants.

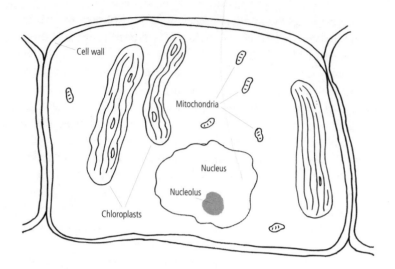

Figure 2.1 Schematic view of a plant cell. Important compartments include: the cell's nucleus, where the DNA is kept; the mitochondria, which carry out the energy metabolism; and the chloroplasts, which are in charge of photosynthesis.

As photosynthesis in plants comes as a pre-packed functional unit, the chloroplast, the question is, where did plants get the chloroplasts from? It is now generally accepted, if not strictly proven, that the membrane-surrounded organelles in the eukaryotic cell are the descendants of formerly free-living bacteria, which entered into a symbiotic relationship with the early eukaryotes, in the course of which they lost most of their genetic repertoire. Lynn Margulis first proposed this explanation for the origin of mitochondria in the late 1960s and since then it has been generalized to other organelles including chloroplasts.

According to the Belgian biochemist Christian de Duve, who won a Nobel prize for discovering one kind of organelle (the liposomes), the ancestral eukaryote must have started out by losing the rigid cellwall characteristic of bacteria, so that it could grow to bigger sizes. This blown-up bacterium would probably have relied on organic nutrients produced by other bacteria (i.e. it would have been heterotrophic, as opposed to an autotrophic organism that is independent of such nutrients). One way in which its supply system may have worked is still demonstrated today by the so-called stromatoliths, which are layer-cakes of different bacterial species: in these, a primitive, autotrophic species feeds the layer of heterotrophs.

Fossilized stromatoliths revealing this layer structure have been dated to ages of up to 3.5 billion years, suggesting that the ancestors of eukaryotes might have started out in such an arrangement with the precursors of chloroplasts. In such a protected environment with a guaranteed food supply, the heterotroph could have lost the rigid bacterial cell wall and developed its membrane into a food import system that would have been able to import bigger lumps, and maybe even whole bacteria. For this purpose it may have developed the trick of engulfing the food with a pocket of its membrane, which then separated from the membrane and ended up as a globule within the cell. This process, known as endocytosis, is still in use in certain kinds of eukaryotic cells, ranging from free-living protists to the killer cells of our immune system.

It's exactly this process of swallowing a whole bacterium which allows us to give a plausible explanation of how the compartments of the eukaryotic cell came into being. By swallowing photosynthetically active bacteria, which in exceptional cases were not digested but carried on their usual business, the early eukaryotes acquired chloroplasts. Similarly, engulfing bacteria with other specific abilities resulted in the other organelles, such as mitochondria, peroxisomes, etc. De Duve's scenario is corroborated by the observation that all the internal membranes of the eukaryotic cell show a closer resemblance to bacterial membranes than to the outer eukaryotic cell membrane.

Note that the 'invention' of the eukaryotic cell was a major breakthrough enabling the evolution of multicellular organisms. Without it, life on Earth might have remained exclusively microbial to this day, as it had been for the preceding two billion and a few years.

We now know where eukaryotes, including today's plants, got photosynthesis from—much like a big business swallows a smaller, more specialized one—but this also means that there was photosynthesis before the complex cell came into being, and that we need to trace its roots further back in time. To be able to do that, however, we need to discuss some basic elements of how photosynthesis actually works.

How plant photosynthesis works: a rough guide

Photosynthesis, although arguably the most important biochemical reaction for life on Earth, and the source of almost all our energy (including fossil fuels) and carbon nutrients, has proved a hard problem for scientists over the last four centuries. One of the reasons why it was only discovered quite recently is that one of the reagents, carbon dioxide, is an invisible and relatively sparse gas in the air. Thus the crucial experiment to address the question of where plants get their food from had already been carried out more than 350 years ago by the Belgian physician Johan-Baptista van Helmont (1579–1644). His experiment was so well thought through that it is quoted by historians of science as one of the cornerstones of modern experimental biology. And yet he came to the wrong conclusion, because he could not know about carbon dioxide. His failure to recognize that plants take up a gas from the air is ironic, because Helmont was the very person who coined the word 'gas' (which, like the English word 'chaos', derived from the Greek *khaos*). The chemical identification of gases, however, was only accomplished in the eighteenth century, by the chemists Lavoisier and Priestley.

Helmont planted a young willow tree in a pot, carefully weighed the whole ensemble, and made meticulous records of any inputs and outputs. After five years, the tree had gained more than 70 kilograms in weight, without any significant loss in the weight of the earth in the pot. Helmont came to the conclusion that the mass increase came from the water. Although he had given gases their general name, he believed air to be a fundamentally inert medium, which is why he never considered the possibility that it might take part in the metabolism of his willow tree.

A century later, the British chemist Joseph Priestley (1733–1804) realized that plants do in fact metabolize atmospheric gases, and that they do it in exactly the reverse direction to animals and combustion. In the course of his fundamental research into the chemistry of air, Priestley showed that air 'used up' by a burning candle can be 'regenerated' by a growing plant. Towards the end of the eighteenth century it was believed that plants cleave carbon dioxide into carbon and oxygen, keeping the former and excreting the latter. As the elemental composition of sugars and starch corresponds to one water molecule per carbon atom, the name 'carbohydrate' was coined on the somewhat naïve assumption that these compounds form by the addition of water to carbon atoms resulting from the cleavage of carbon dioxide.

This hypothesis can explain the turnover of atoms in photosynthesis with the correct numeric relationship ($n\,CO_2 + n\,H_2O +$ sunlight $\rightarrow (CH_2O)n + n\,O_2$), and hence it withstood all the scrutiny that nineteenth-century chemists could subject it to. One and a half centuries were to go by before it could be shown that this interpretation was actually wrong. Only the availability of radioactively labelled carbon compounds after the Second World War enabled researchers to follow the reactions in detail and to work out which atom goes where (see FLASHBACK: Melvin Calvin, p. 40). It turned out that the oxygen produced by plants does not come from the carbon dioxide, but from the water. While this may look like a pedantic different-iation to make, it is in fact a breakthrough discovery that enabled the thorough understanding of the chemical mechanism of photosynthesis.

Today we know that the reaction which splits the water molecule and harnesses the energy of the sunlight (the light reaction of plant photosynthesis) is handled separately from the one which uses the carbon dioxide (the dark reaction). Most plants separate the two processes in space, but some tropical plants run them at different times. Cacti and other plants adapted to drought close their pores during the daytime to minimize the loss of water through evaporation, so that they can only take in carbon dioxide during the night. Such day/night cycles are typically controlled by biological clocks, which we will explore in Chapter 4.

The photosynthesis apparatus of plants is one of the most complex molecular machines, including more than a dozen molecular components which have to be arranged in a particular way so that the electrons can travel from one to the other. For the purposes of this book, it is sufficient to know that the process can be conceptually split into two parts, one of which requires light, while the other doesn't. The light reaction converts the energy of the sunlight into chemical energy. This is stored in small molecules (ATP and NADPH). During the dark reaction, this chemical energy is harnessed for the construction of organic (carbon-containing) molecules starting from carbon dioxide. The key metabolic pathway in this process is a cyclic process known as the Calvin cycle (see FLASH-BACK: Melvin Calvin). The key enzyme of the dark reaction, a protein named rubisco (short for ribulose-1,5-bisphosphate carboxylase), accounts for roughly 30 per cent of the protein content of plant leaves, and is thus the most abundant protein on our planet. The regulation of its activity is a main determinant of the speed of plant growth.

There is one more complication which needs to be told, however. Chloroplasts in plant cells use a combination of two different kinds of 'solar panels' to catch the light (Figure 2.2). These are known as photosystems I and II (PS I and II), and they work in a sequential arrangement first proposed by Hill and Bendall in 1960. Photosystem I is presumably the older one, as it can work (as it does in certain bacteria) independently of photosystem II and uses solar power to produce organic chemicals. This simplified version of

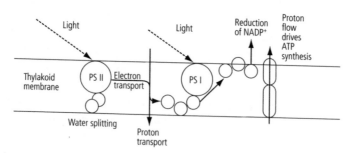

Figure 2.2 Schematic overview of the most important functional elements in plant photosynthesis. Electrons obtained from the water-splitting reaction (bottom left) are used to reduce NADP+ (top right). Protons are actively transported into the cell. When they flow back out, they are used to provide energy for the synthesis of ATP.

photosynthesis, however, does not include water splitting or the production of oxygen. While plant photosynthesis is quite uniform, different groups of bacteria do all kinds of things with sunlight, as we shall find out in the next section.

From sunbathing bacteria to chloroplasts

Now we can return to the question raised previously as to where the ancestors of modern-day plants got chloroplasts from. From bacteria, presumably, but there are many different kinds of photosynthetic bacteria.

The simplest and therefore also the best-understood approach to harnessing sunlight to create chemical energy is that of the salt-loving archaea of the genus *Halobacterium*. Up to 50 per cent of the outer cell surface of these microbes can consist of the purple-coloured patches known as the purple membrane. This can be isolated as a pure sheet and represents a quite remarkable material. Its main component is a protein, bacteriorhodopsin, which essentially collects sunlight and uses the energy to pump hydrogen ions (protons) out of the cell (Figure 2.3(a)). Thereby it creates a gradient in the concentration of these ions, which is a kind of energy storage. Other proteins that allow the protons to flow back in can use the gradient to generate highly energized molecules known as ATP.

However, this kind of solar energy conversion does not result directly in the formation of new carbon–carbon bonds, so one wouldn't normally call it photosynthesis. There are five major groups of bacteria which are photosynthetically active in a strict sense. Four of these have only got one photosystem (take a deep breath before reading these names): the Chlorobiaceae (green sulphur bacteria, e.g. *Chlorobium*), the Chloroflexaceae (green non-sulphur bacteria, e.g. *Chloroflexus*), the Thiorhodaceae (purple sulphur bacteria, e.g. *Chromatium*), and the Athiorhodaceae (purple non-sulphur bacteria, e.g. *Rhodospirillium*). The latter group also includes the purple bacterium *Rhodopseudomonas viridis*, which Robert Huber, Johann Deisenhofer, and Hartmut Michel used for their pioneering determination of the first ever crystal structure of

Melvin Calvin and the mechanisms of photosynthesis

THE ESSENCE OF BIOCHEMISTRY can be reduced to a complex network of lots of metabolic reactions: a gigantic street map with main and side roads, cul-de-sacs, frontiers, and roundabouts. One important roundabout is in charge of burning carbon-based nutrients to carbon dioxide, which happens in the mitochondria and is known as the Krebs cycle. In stunning symmetry, the reverse process, namely the incorporation of carbon dioxide into carbohydrates using the energy of photosynthesis, is also a roundabout. In its key step, a sugar molecule with five carbon atoms (ribulose-1,5-bisphosphate) reacts with the carbon dioxide, resulting in a molecule which is first split in two, then reassembled to a diphosphate of the sugar fructose, which forms one half of ordinary household sugar and allows the plant to form other carbohydrates such as glucose, starch, etc. This cycle, which looks quite scary if you write down all the formulae but is really just a means of turning carbon dioxide into carbohydrates, is named after the chemist Melvin Calvin (1911–97).

Just after the Second World War, Calvin, the son of Russian immigrants in Minnesota, turned out to be the right person in the right place at the right time to crack the mystery of photosynthesis. The radioactive variant of carbon, the ^{14}C isotope, had only been discovered in 1940 at the radiation laboratory of the University of California at Berkeley. A project to use this isotope for photosynthesis research had been started, but then abandoned during the war. After the war, the founder of the laboratory, Ernest Lawrence, encouraged Calvin to use this isotope, which was now available in reasonable amounts, to have another go at photosynthesis.

Calvin and his coworkers at the Old Radiation Laboratory in Berkeley grew cultures of the alga *Chlorella*, which at a certain short time interval (the pulse) they fed with radioactive carbon dioxide. Then they applied various analytical techniques, most notably paper chromatography, to find out which substances contained the labelled carbon at which times after the pulse. (To try the principle of paper chromatography without the radioactivity, dip one end of a piece of blotting paper in any coloured liquid, and observe how different components migrate at different speeds.)

When Calvin's coworkers carried out the experiment as quickly as possible, applying a pulse of only five seconds, and killing the cells immediately afterwards by suspending them in alcohol, the paper chromatogram of the cell extract only contained one radioactive spot, which they could identify as 3-phospho-glycerate. Although this compound appears almost instantaneously, it is not the first product of carbon dioxide incorporation. It is only one half of an unstable compound of six carbon atoms, which arises from the reaction between the five-carbon compound ribulose diphosphate and carbon dioxide.

By varying the pulse and subsequent reaction times, backed up by painstaking comparative studies to identify the radioactive spots obtained in the chromatograms, Calvin and his coworkers could elucidate the metabolic pathways of photosynthesis by the end of the 1950s. The last pieces of the jigsaw fell into place, as Calvin was to reminisce later on, when he was sitting at the wheel of his car waiting for his wife. After months of not being able to fit together the latest results, a sudden inspiration allowed him to piece the complete picture together within 30 seconds. Only a few years later, in 1961, he received the Nobel prize in chemistry for this work.

From the early 1950s onwards, Calvin was also involved in research addressing the origin of life. In an experiment similar to, but pre-dating, the famous Urey–Miller experiment, he investigated whether radioactivity could make prebiotic 'soups' come up with organic molecules, but didn't get more than formaldehyde and formic acid out of it.

Calvin remained in Berkeley through to his retirement, working on a host of projects, including the chemical basis of memory, cancer, and the use of plant oils as fuel.

(a)

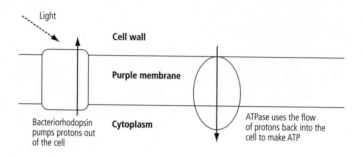

(b)

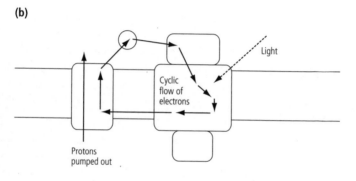

Figure 2.3 Comparison of the functional elements and net chemical reactions in different kinds of bacterial photosynthesis: (a) the purple membrane of halobacteria; (b) the photosynthetic reaction centre of *Rhodopseudomonas viridis*.

the complex molecular apparatus which converts light to chemical energy (the photosynthetic reaction centre, shown somewhat over-simplified in Figure 3.3(b)), which earned them the 1988 Nobel prize. For the first time a structural basis became available for the detailed analysis of the molecular processes in photosynthesis. Somewhat surprisingly, the results of this study also suggested an evolutionary relationship between the *R. viridis* system and the eukaryotic photosystem II. More generally, it has been confirmed that the photosystems from purple bacteria resemble PS II, while those of green bacteria are of the PS I type.

The discovery and detailed investigation of green and purple photosynthetic bacteria in the 1930s brought important conceptual insights, as these bacteria use analogous but different reactions to convert carbon dioxide into carbohydrates. While plants take the hydrogen for this reaction from water (leaving oxygen as the waste product which they ruthlessly emit into the atmosphere), sulphur bacteria take it from hydrogen sulphide, leaving elemental sulphur as waste. The non-sulphur bacteria can use various small organic molecules, resulting, for instance, in the conversion of isopropanol to acetone. For the chemist, photosynthesis is essentially a redox reaction (i.e. electrons are shifted from one atom to another one), in which carbon dioxide is reduced (i.e. the carbon gets more electrons in its new position in a carbohydrate) while some other compound (water, hydrogen sulphide, or isopropanol) is oxidized (loses electrons).

The biggest, most successful, and most sophisticated group of photosynthetically active bacteria is that of the cyanobacteria. As these organisms are often found to form relatively complex multicellular structures resembling algae, they were originally classified by botanists as 'blue-green algae'. Only in the 1970s could bacteriologists persuade plant scientists that these 'algae' should really be seen as bacteria, and since then the new denomination 'cyanobacteria' has found widespread acceptance. Apart from the fact that they clearly lack a nucleus, molecular analysis of their genetic heritage has firmly placed them in the family tree of the bacteria (the first genome sequence of a cyanobacterium, *Synechocystis*, was published in 1996).

Apart from their ability to form communities almost matching the complexity of a multicellular organism, cyanobacteria are remarkable for their resistance to extreme and changeable environmental conditions, including high temperatures, drought, salinity, alkaline pH, and very high or low light intensity. This, in combination with the fact that they are independent of organic nutrients because they can get both carbon and nitrogen from the air, enables them to live just about everywhere. They can be found in the cracks of rocks in the Namibian desert, in tidal zones, in the Antarctic, and in salt lakes as well as in freshwater lakes. Strongly acidic pH appears to be the only environmental stress condition that other species can cope with but cyanobacteria cannot. Among the cyanobacteria we also find the only known examples of bacteria possessing an independent biological clock, as will be demonstrated in Chapter 4.

Cyanobacteria are also very likely to be the closest living relatives of those bacteria which turned into chloroplasts, because they are the only non-eukaryotic organisms that possess a two-step photosynthesis mechanism. They are also the only bacteria to use the green pigment chlorophyll *a* (a chemical similar to the red pigment heme which we carry in our red blood cells, but with magnesium rather than iron in the central binding site) as an antenna serving both photosystems in much the same way as the chloroplasts of modern plants do. And they are still very keen to engage in symbiotic relationships, e.g. pairing up with fungi to form lichens.

From fossils we know that 3.3 billion years ago there were microbes which resembled today's cyanobacteria, and chemical evidence suggests that some photosynthesis took place at that time. The fossilized stromatoliths mentioned above suggest that close communities between photosynthetic and heterotrophic microbes existed in the very early days of cellular life. Obviously the fossilized shape does not tell us which—if any—photosystem these microbes possessed, but many scientists believe that the crucial evolutionary step from the primitive, single-step photosystem I photosynthesis to the complex mechanism used by today's plants first occurred in these bacteria. So they probably invented the photosynthesis reaction which splits the water molecule and leads to the production of molecular oxygen. And in making this step they triggered the biggest environmental catastrophe in the history of our planet.

The mother of all environmental catastrophes: global change triggered by photosynthesis

When our planet first formed, gases seeping out of the newly accreted hot material formed the first atmosphere—but its composition remains unknown to us. We can, however, analyse some of the oldest known minerals (current record-holders are zircons from Western Australia, believed to be 4.4 billion years old) and conclude from their chemical composition what the atmosphere may have been like just a few hundred million years after planet Earth was born.

From such studies it is well established that, before the environment was dominated by the presence of life, the chemical properties of the atmosphere were not dominated by the highly reactive and oxidizing oxygen gas that we need to survive. Apart from chemically inert gases such as nitrogen and carbon dioxide, the early atmosphere also contained reducing gases such as ammonia and methane, and minerals exposed on the surface also provided reducing potential. In such an environment, animals, which get their energy from the combustion of organic food, could not have even started to evolve. Within the first 1.5 billion years of Earth's history, the reducing gases vanished from the atmosphere, as a result of both geochemical and the first biological processes. If, at that stage, oxygen-producing cyanobacteria were already around, the oxygen would have immediately reacted with the reducing gases and minerals, forming neutral and inert compounds such as water and carbon dioxide.

Some 2.5 billion years ago, there were no significant amounts of reducing substances left in the atmosphere. Primeval cyanobacteria were widespread and busy producing oxygen, which now no longer reacted with reducing agents, but accumulated in the atmosphere, which therefore became an oxidizing environment. Note that this change not only affected the composition of the atmosphere but also turned many of the geological processes upside down, making a number of elements occur in a different chemical form than before. For instance, in a reducing atmosphere you would expect to find iron as a pure, shining metal or as soluble iron salts (oxidation state II) rather than as the rusty nails and iron ores (oxidation state III) that we know today.

And the changes in the geochemistry in turn affected the living organisms at least as much as the changing atmosphere. Among the elements and compounds important to life today, copper, zinc, chloride, bromide, iodide, and the oxides of sulphur, nitrogen, vanadium, and molybdenum became more available to organisms as a consequence of the advent of oxygen. On the other hand, molecular hydrogen and iron oxides became less available. The Oxford chemist R. J. P. Williams has suggested that this change in the availability of chemical elements and molecular building blocks may have made an important contribution to the subsequent evolution of multicellular organisms.

In the course of roughly one billion years the oxygen content in the atmosphere gradually increased to the level of about 21 per cent, where it has stagnated to this day. This value is not coincidental, but is controlled by lots of geochemical and biological processes, making up a global self-regulatory system. Imagine for a moment that the oxygen content of the air were to rise by just a fifth, from 21 to 25 per cent. This change would lower the ignition point of combustible materials quite dangerously. Any pile of dry leaves in the sunshine could ignite spontaneously under these conditions. There would be widespread forest fires which would both use up oxygen by themselves and reduce the vegetation which produces it. Within a few years the value of 21 per cent would be re-established.

At the time when the oxygen produced by cyanobacteria started to pollute the previously oxygen-free atmosphere, there were no organisms around which would have been able to cope with an oxidizing atmosphere. Although there were many different species of bacteria, they would have all been in the class now known as anaerobes: oxygen would have been a poison for them. However, as this poisoning of the atmosphere was drawn out over a very long timespan (one billion years as compared to the few hours' generation span of typical bacteria), the existing species had plenty of time to either adapt or escape into environments which remained free of oxygen.

Among those that adapted and learned to use oxygen to burn things were those bacteria which eventually ended up as mitochondria in the eukaryotic cells. Their achievement is almost as important for the further development of the biosphere as the development of oxygen production, because it allowed the chemical circles of oxygen and carbon dioxide to be closed in a self-regulating system which keeps the atmospheric concentrations of these gases constant at a value which is favourable for our flourishing biosphere. Essentially, this circle of life, which is today most visibly populated by plants and animals, was originally invented by the bacterial ancestors of chloroplasts and mitochondria.

Now, a couple of billion years later, there are also lots of free-living microbes around, which can either tolerate or even use oxygen. Those that did not adapt but escaped are today found underground, in environments where reducing compounds keep the oxygen at bay. Above ground, they can also be found within cows—where they are the world's biggest producers of the reducing gas methane.

Thus we (and all other animals) owe our existence to this earliest example of global pollution. In a sense we are descendants of the cleanup force that evolution sent out to mop up the oxygen which photosynthesis produced. But the water-splitting revolution not only gave us the oxygen that we breathe—an almost equally important outcome of this development is the presence of the protective ozone layer in the stratosphere.

The endangered shield: the stratospheric ozone layer

Light is what we make of it. There is no fundamental difference between the light of a lamp, X-rays, and microwaves, except for the wavelength. Out of the wide range of electromagnetic waves that the Sun sends our way, we only perceive a very small band of wavelengths. This range, known as the visible spectrum and demonstrated in the rainbow or with a glass prism, reaches from blue-violet to red, corresponding to wavelengths from about 400 to 750 nanometres. Radiation beyond 750 nanometres would be perceived as warmth only (infrared), while we have no way of sensing the radiation below 400 nanometres, which is called ultraviolet or UV light. (We can only see UV when it is absorbed by substances that can convert its energy to longer wavelengths and thus fluoresce in the visible range. This is how brighteners used in washing powders manage to make your clothes look 'whiter than white'.)

The reason why we tend to call different parts of the electromagnetic spectrum by different names is that they will have different effects on things, and especially on living things. As the energy content of electromagnetic radiation increases towards shorter wavelengths, UV light can carry more energy than visible light, and therefore it can be very dangerous for living things. Such high-energy radiation can rip certain molecules apart and thereby start unusual chemical reactions. Within the living cell nucleic acids are most at risk, as all their building blocks absorb UV light (while only 3 out of the 20 amino acids making up proteins absorb UV). Thus life based on nucleic acids could not exist in the UV radiation density that the Sun shines down on the surface of our planet. (In the oceans it was protected by the water which absorbs UV quite efficiently.) Luckily, there is a special protection against it, which life on Earth built for itself as a byproduct of the oxygen produced by photosynthesis.

This protection is known as the ozone layer, although strictly speaking it is not a layer but only a relative enrichment of this otherwise very rare chemical. If the atmospheric ozone was concentrated in a layer and subjected to the atmospheric pressure found at ground level, it would be only a few millimetres thick. Ozone is a less stable (i.e. richer in energy) variant of oxygen. Instead of the two atoms of oxygen contained in a normal oxygen molecule, ozone has three. It can form wherever some energy source (such as UV light) splits oxygen molecules into atoms, so that a free oxygen atom can combine with another oxygen molecule. Because of its high energy content, however, the ozone molecule disintegrates quite quickly when it either absorbs UV light or encounters another molecule that can react with the excess oxygen atom. Therefore, its concentration in the atmosphere remains low, accounting for only 1 molecule in 10 million.

Most of our atmospheric ozone is found at altitudes of 10 to 30 kilometres, which is the lower stratosphere (see Figure 2.4). More precisely, the ozone is responsible for the fact that we have a stratosphere, while Mars and Venus don't have one. The UV energy

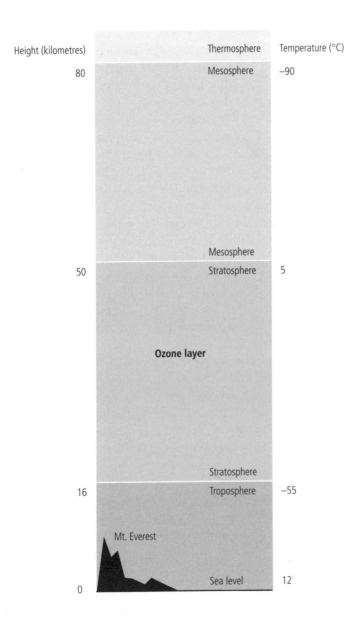

Figure 2.4 Cross-section of our atmosphere indicating the location of the stratosphere and the ozone shield. The thermosphere extends to about 500 kilometres above sea level, followed by the exosphere which gradually fades into empty space.

it absorbs ultimately heats up this part of the atmosphere, leading to an anomalously high temperature layer above cooler parts of the atmosphere. This 'stratification' is what defines the stratosphere and is the reason behind its name. Up there the ozone serves as a shield against the most devastating kinds of UV radiation from the Sun. At ground level, if it forms as a consequence of pollution or of technical use of UV radiation, it is undesirable because of its aggressive chemical behaviour. Most people will know its smell, however, as small quantities are also produced by electrical sparks, for instance in model railways, and by the UV lamps in a solarium if they illegally radiate at too-short wavelengths.

First indications that the ozone content of the atmosphere might be at risk from the effects of human activities were obtained by the Dutch chemist Paul Crutzen, who was then at Oxford and later became a director at the Max Planck Institute for chemistry in Mainz, Germany. He showed in 1970 that the oxides of nitrogen (NO and NO_2, collectively known as NO_x) can catalyse the disintegration of ozone. These compounds are produced in any kind of combustion process using air—their presence in car exhausts, for instance, is a major issue in pollution control. What might become more troublesome for the ozone shield, however, is that these oxides are also produced and emitted at high altitudes by jet engines. The risk would have been even higher if the planned introduction of a large number of supersonic passenger aircraft had gone ahead following the small prototype series of Concordes built in the late 1960s. This project, however, was shelved for economic rather than environment reasons.

While the oxides of nitrogen already had all it takes to play the role of the bad guys, Mario Molina and F. Sherwood Rowland caught a group of chemicals in the act of endangering the ozone shield that had for 50 years been cast as perfect angels. Chlorofluorocarbons (CFCs) were developed at the end of the 1920s as the perfect cooling medium, to replace the toxic and smelly sulphur dioxide, and the equally disagreeable ammonia. CFCs are chemically inert, not inflammable, and non-toxic. It was their remarkable stability in the environment that turned to their disadvantage.

Two links in the fatal chain of events leading from the release of

CFCs to the destruction of ozone had been recognized by other researchers. In the 1960s, the British chemist James Lovelock had invented the electron capture detector, an extremely sensitive method for the detection of such molecules, and had used it to show that CFCs had already spread out over the entire globe, although they were mainly produced and used in the industrialized world. Richard Stolarski and Ralph Cicerone had shown that free chlorine atoms can propagate the destruction of ozone in much the same way as oxides of nitrogen. Molina and Rowland combined these observations with their own findings and described a hypothetical chain reaction: CFCs are produced, serve as coolants for a while, get released into the atmosphere at some point, spread globally, and also rise to the higher levels of the atmosphere, including the ozone-rich stratosphere. Their remarkable stability has turned into a disadvantage here—other kinds of organic molecules would never reach the stratosphere but rather would be turned into carbon dioxide and water vapour on their way. At stratospheric heights, the intense short-wavelength UV radiation splits chlorine ions from the CFCs, which then catalyse the disintegration of ozone molecules.

This theory was published in *Nature* in 1974 and was at first highly controversial. A few years later, however, it could be confirmed by measurements and all that remained to be discussed was the extent of the predicted damage. Today we know that Molina and Rowland had not been quite pessimistic enough—they underestimated the amounts of ozone that would be lost. Crutzen, Molina, and Rowland received the Nobel prize in chemistry in 1995 for the anticipation of the chemical reactions that eventually led to what we now know as the ozone hole.

In spite of these warnings, the real menace of ozone depletion was only brought home to many in 1985 when some worrying measurements by the British Antarctic Survey were published. According to these results, the thickness of the (hypothetically compressed) ozone 'layer' during the Antarctic spring had shrunk from more than three millimetres in the 1960s to less than two millimetres in the 1980s. The Antarctic 'ozone hole' was born, and year after year new measurements showed that the effect was growing in severity, increasing the pressure both on scientists to come up with explanations and possible cures, and on politicians to take action. In the winter of 1995–96 ozone losses were detected in the North Polar region as well.

Scientists investigating why the ozone was most notably disappearing above Antarctica, and why the effect was more pronounced in spring, found that stratospheric clouds which can only form below –80°C, i.e. only in the polar regions, play a crucial role. In the absence of these clouds, the chlorine split from the CFCs can be bound into relatively stable compounds such as hydrogen chloride or chlorine nitrate, which in this context are called the chlorine reservoir. Only with the help of microscopically small particles which are contained in the stratospheric clouds can the chlorine reservoir spring into action. These particles contain nitric acid molecules in an as-yet elusive arrangement with water molecules.

The clouds catalyse the release of the dichlorine molecule, which is split into two chlorine atoms by the UV radiation. These free atoms are a very reactive species and embark on a fatal reaction cycle. Each can react with an ozone molecule to form chlorine monoxide and normal oxygen. Chlorine monoxide in turn can combine with one of its kind to form dichlorine dioxide, which then splits into two chlorine atoms and dioxygen molecules, at which point each of the chlorine atoms can start the circle again and destroy another ozone molecule, and another.

The finding that this reaction mainly occurs above the Antarctic (and to a lesser extent over the Arctic region) does not imply that penguins and polar bears are the only ones put at risk by the ozone hole. It is generally believed that ozone depletion will spread across the globe in the long term. Even with the political measures taken from the late 1980s and the worldwide ban on certain ozone-damaging chemicals in 1996–97 (in an amendment to the Montreal protocol), the depletion will continue, as the CFCs already released into the atmosphere make their slow progress around the globe and up to the stratosphere. A trend reversal is only expected to begin 10 years after the ban; a complete recovery of the ozone layer may take up to a century. However, if Crutzen, Rowland, and Molina had not recognized these problems in the 1970s, things could have become much worse.

Feed the world:
the importance of photosynthesis today

The ozone shield which allows us to live on dry land is just one of many riches we owe to the photosynthesis of cyanobacteria and plants. We breathe the oxygen produced by plants; we feed directly or indirectly on the carbohydrates, proteins, and vitamins that they have synthesized using the energy coming from the Sun. The vast majority of all species populating our planet depend on photosynthesis in one way or another. It has been estimated that this 'oxygen revolution' has increased the global biomass production by two to three orders of magnitude. Apart from being a key element of biology, turning light into chemical energy could also be very useful in technology, if we could learn from nature how to do it efficiently.

No wonder that researchers around the globe are still keen to improve the understanding of the process, and ultimately to bring it to a point where we can mimic it. Although the major features of photosynthesis are clear, and detailed structures of both photosystems have been published, some subtle details of the primary process of catching a photon and turning it into a charge separation are still poorly understood. Moreover, the details of how the complex machinery of the photosynthesis apparatus assembles within a cell and how the parts are fixed in their respective places are far from clear.

Harald Paulsen and his coworkers at the University of Mainz are using genetic engineering to introduce subtle changes into the natural light-harvesting molecules. Apart from addressing questions of how the photosynthesis system originates in the cell, they are also trying to assemble these modified parts into an artificial mimic that could both help to understand the functions and lead to applications in bio- or nanotechnology. At the moment, this research is at the stage of finding the right combinations of elements and suitable substrate materials, which will optimally harvest the light energy and efficiently convert it into electrical currents.

Other approaches to 'artificial photosynthesis' have been reported by several groups. Thomas Moore and his coworkers at

Arizona State University have, for instance, created minute bubbles (vesicles) with a diameter of only around 100 nanometres, surrounded by lipid bilayer membranes similar to the cell membrane. Into this membrane they implanted an artificial organic molecule, called CPQ because it consists of a *c*arotinoid, a *p*orphyrin, and a *q*uinone, oriented such that the latter is at the outer surface of the membrane. They were able to show that, much like in real photosynthesis, the arrival of a photon (coming from a laser, not from the Sun) created a charge separation, converting CPQ into C^+PQ^-. A further quinone compound inserted into the membrane could then take over the negative charge from CPQ, combined with a hydrogen ion from the surrounding solution, and shuttle the two to the inside of the bubble, where the electron would neutralize the charge of the carotinoid and the hydrogen ion would lead to a measurable change in pH.

While Moore's system is strongly inspired by nature (quinones also play important roles in the charge separation in natural photosystems), other researchers have designed more artificial systems, involving components such as colloidal titanium dioxide films (the Grätzel cell), polystyrene modified with ruthenium complexes, supramolecular assemblies, or even fullerenes.

In a way, the more chemical and not quite so biomimetic energy converters are showing the way that biomimetic efforts have to go. If molecular systems are to compete with the semiconductor solar panels currently available in the shops, developers would need to understand the photosynthesis system well enough to be able to make it simpler and more robust, perhaps even reducing it to the elementary setup of one membrane and two molecules, as in Moore's work. In the case of bacteriorhodopsin—hailed as a promising biomaterial for optical applications since its discovery—the difficult handling has so far thwarted all promises, an experience which also suggests that simplified and user-friendly systems should be developed. There are good reasons to try this, as current improvements in the efficiency and affordability of solar panels are still progressing slowly (even though at least one photovoltaics company in Germany has recently reported profits and there is a feeling that the time for more widespread use has now come).

Meanwhile, until the photovoltaic revolution really happens, sunlight remains our most underused energy resource. But why isn't the biosphere using more of it? One has to bear in mind that plants need other things besides sunlight. And as sunlight is present in vast excess in most places, at least above ground or sea level, the other factors will be more likely to limit the growth and spread of life on Earth. As we have seen in the energy balance (Figure 1.2), photosynthesis only uses 0.023 per cent of the light shining on our planet. So what is stopping us?

One serious limit to life in many places is the availability of liquid water, which is why deserts are deserts. Globally, however, the most limited supply on which living things rely is that of chemically usable nitrogen. This is of course a paradoxical situation, as the air consists of more than three-quarters nitrogen gas, holding a reserve of some 30 million billion tons of the substance. The only trouble is that nitrogen gas consists of extremely stable molecules, which are very reluctant to participate in any chemical reaction. Only a few groups of archaea and bacteria (including the cyanobacteria mentioned above) and, since the early twentieth century, the chemical industry, hold the knowledge of how to turn nitrogen into compounds such as ammonia or nitric acid, which make the element usable for the chemistry of life.

The industrial-scale production of ammonia from nitrogen gas, using the process invented by Fritz Haber and developed by Carl Bosch, began in spring 1914 at Oppau near Ludwigshafen on the Rhine. The global spread of this technology and of nitrogen fertilizers enabled the world population to grow from 1.8 billion to more than 6 billion in the twentieth century. After decades of steady increase in industrial production it has caught up with natural nitrogen fixation. In the 1990s each turned over roughly 100 million tons of nitrogen per year, which also implies that of today's six billion people more than half owe their sheer existence to Haber and Bosch. (I am not sure how to balance this figure against the deaths and suffering caused by the First World War chemical weapons also developed by Haber in the mistaken belief that they would shorten the duration of the war.)

It was a little-noticed event, but, some time around 1990, the global population trend actually turned, so that the annual per-centile increase is now significantly smaller than at the peak in the late 1980s. But the total numbers will, of course, continue to rise for a little while, up to a plateau predicted to be around 9 or 10 billion. If the planet is to feed so many people in the mid-twenty-first century, it will need even more nitrogen than today (unless everybody turned vegetarian, in which case the primary resources would be used much more efficiently!). This will involve problems of both environmental impact and distribution. Getting plants to do the nitrogen fixation for themselves by genetic engineering, although very unpopular today, may become a necessity.

Apart from food, people will also need an ever increasing amount of energy. We have seen in Chapter 1 that global energy consump-tion is dwarfed by the energy the Sun shines down on us—so it would only be a question of using this more efficiently. (Burning fossil fuels is in fact a way of using solar energy, but certainly not the best one.)

This is one area where twentieth-century technology has per-formed rather poorly. Techniques to convert light to electricity have quite appallingly low efficiencies, while biologically inspired approaches are not very economical. Part of the problem is that high demand for energy and high availability of sunshine don't usually occur in the same place (except for California!). Thus converting sunlight collected in the Sahara into a form of energy that can be used in Norway will be one of our major tasks for the twenty-first century, and maybe an improved understanding of photosynthesis would help to achieve this.

CHAPTER 3

Creatures that glow in the dark

OXFORD'S HIGH STREET is where many of the famous colleges are lined up. Among them on the south side of the road is University College, where a student called Stephen Hawking (whom you may remember as a guest star of both *The Simpsons* and *Star Trek*) got his first degree in physics. On a tall beige sandstone wall linking two buildings of this college you will find a rather unspectacular grey slate plaque with the inscription:

In a house on this site
between 1655 and 1668 lived
ROBERT BOYLE
Here he discovered BOYLE'S LAW
and made experiments with an
AIR PUMP designed by his assistant
ROBERT HOOKE
Inventor, Scientist and Architect
who made a MICROSCOPE
and thereby first identified
the LIVING CELL

The plaque does not fully do justice to the importance of the site, because it was here that a rather dramatic experiment (involving the said air pump), which I will describe later in this chapter (so don't change the station!), laid the foundation for the study of bioluminescence. Thus, from a very early achievement of life on Earth that has influenced its development like nothing else, we now move on to a phenomenon that has probably arisen relatively late and appears almost to be a luxury. While photosynthesis was all about turning light into chemical energy, we shall now discuss organisms that do the exact opposite, namely, turn chemical energy into light.

Life in the deep dark sea

If life depends on light, then the deep sea, which is pitch dark beyond a depth of a few hundred metres, must be lifeless. This was at least what most nineteenth-century scientists assumed, with the notable exception of the great-grandfather of science fiction, Jules Verne, who correctly predicted (in *Twenty thousand leagues under the sea*) that life was possible more than 10 kilometres below sea level. But then, in the 1880s, French biologists demonstrated that samples retrieved from depths of up to 5100 metres by the research vessels *Talisman* and *Travailleur* contained living microbes. In contrast to common or garden bacteria these organisms reproduced happily at the high pressures of 500 to 600 atmospheres, corresponding to the depth of their habitat.

The realization that the oceans are indeed habitable at any depth implies that the terrestrial part of the biosphere is only a small fraction of it. More than 60 per cent of the habitable volume lies more than 1000 metres below sea level; it is thus exposed to pressures of more than 100 atmospheres and is virtually devoid of light. It was hard to imagine what kept the bacteria alive so far away from the light of the Sun. Nutrients formed in the sunlit upper parts of the oceans could sink to the bottom and feed organisms there, but this kind of mechanism would only support a very sparsely populated ecosystem which, although not sterile as was previously thought, would qualify as a desert by anybody's standards.

That's why scientists were quite surprised when, nearly a century later, they found extremely rich ecosystems at the bottom of the sea. During a dive with the research submarine *Alvin* in 1977, the geologists John Corliss and John Edmond discovered the first of these biotopes flourishing at a site with warm springs at the geologically active spreading zone of the sea floor near the Galapagos Islands. Two years later, chimneys which eject volcanically heated fluid at around 350°C were discovered. These hydrothermal vents or 'black smokers' are typically surrounded by even more spectacular biotopes, including clams, sea anemones, and, most strikingly, entire forests of the tube worm *Riftia pachyptila* with their characteristic white tubes crowned by bright red gills. Both the biomass density and the species diversity of these areas can compete with sunlit coastal waters, a feat which could not be explained plausibly in terms of marine food chains based on photosynthesis.

It has to be admitted that we still know very little about the sea floor. As scientists involved in its exploration like to point out, the topography of the far side of the Moon is known in more detail than that of the ocean floor. Thus it is perhaps not surprising that the deep sea keeps surprising us. Surprise discoveries like that of the coelacanth in 1938, the black smoker biotopes in the 1970s, and the presence of massive methane hydrate deposits in the 1990s, keep reminding us of how little we know about the deep sea. And what happens down there is important for our life up on the land. The composition of sea water, for instance, is determined not only by what the rivers wash into the sea, but also by the hot springs which remove certain minerals from the water and add in others. The whole volume of our oceans passes through these vents once in eight million years.

Thus it is understandable that scientists from a range of disciplines have eagerly adopted the vents and their biotopes as their field of study. Hundreds of new species have been characterized, and at one point researchers at Santa Barbara, California, even succeeded in growing the notoriously sensitive tube worms in a pressurized tank in the lab. Research into these intriguing sea floor ecosystems soon uncovered that they are in fact independent of the organic nutrients produced by photosynthesis, even though they still benefit from photosynthesis through their use of oxygen. Their food chain is based on chemosynthesis, a reaction in which reduced sulphur compounds such as hydrogen sulphide and metal sulphides are oxidized using oxygen. Much as in photosynthesis, this energy is used to convert carbon dioxide (dissolved in the sea water) into organic molecules which eventually feed the organisms higher up the chain. New forms of symbiosis between primary producers (sulphur bacteria which do the chemosynthesis) and higher organisms such as the clams and tube worms which use these nutrients have been described.

Some intriguing details uncovered by this research showed to what extent life has adapted to this environment. Haemoglobin, for instance, the oxygen carrier in our blood, is easily poisoned by sulphides. Tube worms, in contrast, have to transport both oxygen and sulphides through their blood stream to supply it to their symbiotic bacteria. If the two kinds of molecules got too close to each other in transit, they would react spontaneously, and the energy would be lost as heat, rather than converted into biochemical energy that can be stored. The tube worm has developed a haemoglobin variant which has separate binding locations for both molecules so that they can safely and separately be transported to the trophosome, the 'feed sack' where the sulphur bacteria will use them for chemosynthesis.

But how does the absence of light affect these organisms? Tube worms and some clams are stationary like plants; they don't need eyes any more than a tree does. Some shrimps of vent biotopes have eyes in spite of the permanent darkness they live in. While this puzzled scientists at first, they soon found out that these eyes allow the animals to sense the infrared radiation of the hot water jet of the vent. Considering that this hyperthermal fluid would convert any animal to processed seafood within seconds, this is certainly a useful invention. (As Cindy Lee van Dover has pointed out, *Alvin* pilots tend to share the shrimp's respect for the hot jets, knowing that the windows of their submarine are made of plastics that will melt well below 100°C.)

Deep-sea fish and squid, being mobile and likely to be both predators and prey in the extensive marine food web, are more likely to mind the absence of light for seeing. Hence there are some species which use 'headlights' simply to have the advantage of seeing their prey either in the deep or at night. Many more, however, use light in a variety of other ways to do with attracting mates or prey and repulsing or deceiving enemies, as we shall see below.

The idea that nature had 30 times

Nearly everybody knows fireflies, as there are more than 2000 different species of these luminescent beetles around the world, and many people may have seen luminescent creatures in tidal waters. These are only the best-known examples out of a very wide variety of bioluminescent organisms, which are randomly scattered across all kingdoms of life. Table 3.1 gives a quick guide to some of the better studied examples, but is by no means exhaustive.

Because bioluminescence doesn't make anybody ill, and the few luminescent systems that have found practical applications are well described anyway, bioluminescent creatures are not near the top of the priority list for genome research. (Although some freshwater arthropods can get infected by luminescent bacteria, until 1986 no luminescent bacterium was known that could infect humans. In that year, however, a man who injured himself while cleaning fish was infected with a luminescent strain of a pathogen, *Vibrio vulnificus*, of which he died two weeks later. A few cases of seafood that was observed to glow in the dark were investigated by the US Food and Drug Administration. Contamination of cooked seafood by uncooked material due to bad hygiene practice was normally the cause—which can be more dangerous and difficult to detect if non-luminescent pathogens are involved.) Thus what we know about their evolutionary relationship is still very much based on classical phylogenetic family trees rather than genomes. Bioluminescence is found scattered across an eclectic gathering of organisms with no obvious generalizing principles. It is fairly common in the oceans but almost absent in fresh water. Various groups of worms and insects can glow, but no spiders. Among the vertebrates, fish are the only group using it. From this distribution it appears that there are around 30 non-related bioluminescence systems, which could mean that bioluminescence was invented independently up to 30 times by organisms which had some use for it.

An alternative explanation suggested by William D. McElroy and H. H. Seliger implies that the ancestral luminescence system did not evolve to produce light, but rather to detoxify the oxygen that was introduced into the atmosphere by the cyanobacteria when they

Table 3.1 A selection of bioluminescent organisms

Common name	Systematic name	Habitat	Description and purpose of luminescence
Fish species			
Deep-sea anglerfish	*Melanocoetus johnsoni*	Oceans below 90 m	Fishing rod fin with luminescent bait
Deep-sea hatchet fish	*Argyropelecus aculeatus*	Oceans at 300–500 m depth	Faint blue glow on the underside of the body for camouflage (disrupts silhouette)
Flashlight fish	*Photoblepharon palpebratus*	Oceans at *c.*30 m depth	'Headlights' underneath the eyes serve night hunting
Planktonic organisms			
Dinoflagellates	*Ceratium longipes* and others	Oceans	Light up when disturbed. This optical alarm attracts predators of their predators
Insects			
Firefly	*Photinus* and *Photuris* genera	North America (other firefly species found in warm climates around the world)	Mating signals; *Photuris* females also use it to prey on *Photinus* males
Railroad worm (not really a worm but a beetle!)	*Phrixothrix* species	South and Central America	Bright red headlight and yellow lights along the body sections, used to warn attackers, also when attacking or mating
Fungi			
Honey fungus	*Armillaria mellea*	Attacks dead trees	Threads emit greenish glow, benefit unknown

invented two-step photosynthesis (see Chapter 2). It is indeed striking to note that with all the differences in detail, most biolumi-nescence systems known so far use oxygen (all luciferase-based systems use oxygen but the *Aequorea victoria* system (p. 73) does not!), so maybe some early bacterium evolved a mechanism that disposed of oxygen by turning its chemical energy into light. Later on, when mitochondria allowed better use to be made of this energy, there wasn't much point in keeping this mechanism, except if you had any use for the light produced. So the genes involved, not being much use for many of the descendants of the primeval bac-terium that had them, were likely to get lost or at least mutated quite easily. Later, some organisms may have revived them when they moved into habitats where having light actually was some use.

At this point, we cannot distinguish between these possibilities. Genome information from the organisms involved would, however, enable us to tell whether we really have separate inventions here or rather the scattered descendants of a set of genes that went through a period of being quite unfashionable.

Either way, let's look at how these creatures produce light, to get an impression of the amazing variability of this phenomenon. If you are a higher organism and want to glow in the dark, you basically have two options—either you find some luminescent bacteria and give them shelter, food, and oxygen, so they will produce light for you, or you evolve some kind of gland structure which allows you to mix the reagents for the light reaction yourself.

Luminescent bacteria don't have a light switch; they glow as long as they have fuel to burn. So if you want to be able to switch the light on and off, you either have to suppress their food supply or invent some shutter mechanism. Two shallow-water fish species with bacterial 'headlights' underneath their eyes, both called 'flashlight fish', have developed intriguingly different solutions for this prob-lem. *Anomalops katoptron* has an eyeball-like light organ with a reflective layer at the back. To switch it off, the fish can rotate the ball by 180 degrees, so that the reflective coating serves as a shutter. The other flashlight fish, *Photoblepharon*, in contrast, has a stationary light organ, with a mobile shutter that covers it up like an eyelid.

The majority of luminescent animals, however, prefer to have better control of their light production and thus have evolved glands that secrete the reagents. The control may be mediated either by hormones or by nerves. The first would typically be the case in animals that glow or flash constantly in certain life phases (e.g. during the mating period) or in diurnal rhythms. Neuronal control is of course more specific and can be timed to a precision in the millisecond range.

Fireflies (known as glow worms in some parts, although they are positively insects, not worms) are well-studied examples of this kind of signalling. At night the male flies around lighting up its lantern in a series of short flashes (a few hundred milliseconds) to attract the female on the ground, which then signals back to express her readiness. Starting with the classic works of William McElroy (FLASHBACK, p. 70) in the 1940s and 1950s, biochemists worked out the light reaction which will be described in the next section, and have identified a nerve signal that controls the flashes.

But, as the nerve endings are some 17 micrometres away from the cells which produce the light, it was unclear how the signal bridges this gap. Barry Trimmer and his coworkers at Tufts University (Medford, Massachusetts) have now found that, for a fulfilling love life, the firefly depends on the same signalling molecule as the male of our own species: nitric oxide (NO).

Trimmer had previously studied the role of NO in caterpillars, but after hearing a talk on fireflies decided to check whether they use this molecule too. The fact that they produce nitric oxide synthase in the cells of their lantern suggested they might use it. When his group locked up fireflies in an atmosphere containing some 70 ppm of this gas, they could observe the insects glowing permanently. In order to demonstrate that the NO is acting on the light organ rather than on the brain, they also prepared animals whose nerve connections to the light organ were interrupted. While these were unable to flash spontaneously, the exposure to NO again resulted in rapid flashing or even permanent light. Conversely, when they stimulated the lantern by supplying the nerve signal, they could suppress the light emission by adding NO scavengers.

Thus there is clear evidence that NO is involved as a messenger in the final steps of the signalling. Quite how it fits into the chain of events remains to be clarified, but Trimmer speculates that it may inhibit the respiration process in the light-emitting cells, thus making available oxygen that would otherwise have been used up by the mitochondria. This is consistent with the observation that other inhibitors of respiration (like cyanide) also induce the luminescence. This also suggests a mechanism for terminating the short flashes: as white light suppresses the effect of NO on respiration in mammals, the firefly's light might literally turn itself off.

Some animals, and fish in particular, have not contented themselves with inventing the light bulb, but have built veritable optical equipment around it, including lenses, mirrors, and light guides. Now that we have considered the various bulb and lamp designs, let's find out how they work.

Just pour and watch it glow

In the evening of 29 October 1667 the Irish chemist and physicist Robert Boyle (1627–91) and his assistant dressed up in 'black Cloaths and Hats' and removed the candles from their laboratory at Oxford (which sadly has not survived, but is commemorated by the slate plaque mentioned at the beginning of this chapter) for a spooky experiment which required absolute darkness. They were about to use the vacuum pump ('Pneumatick Engine') which Boyle and Hooke had built in 1659 after they heard of the vacuum experiments of Otto Guericke in Magdeburg. In the receiver of the pump they placed a piece of 'shining Wood'—dead wood infested with bioluminescent fungi. As they removed the air from the vessel containing the wood, they observed that the luminescence faded more and more. After the tenth suction, they could perceive no more light at all. In contrast, when they allowed the air back in, the light returned 'almost like a little Flash of Lightning'. Repetitions proved that the luminescence also returns if the vacuum is maintained for up to half an hour. Similar results were also obtained with a luminescent sample of 'stinking Fish'.

While all these observations make perfect sense with the benefit of our present-day biochemical knowledge, they must have been quite mysterious for Boyle. In the communication ('New Experiments Concerning the Relation between Light and Air') which he sent to the Royal Society the very next day, he only hints at the philosophical implications regarding the nature of life, light, and air, which others might derive from these experiments. Today, of course, we remember this experiment because it was the first step towards the identification of one of the four most important reagents in bioluminescence: oxygen.

Two of the others were identified in 1885, in a major breakthrough on which all investigations of the physiology of bioluminescence are based, by the French physiologist Raphael Dubois (see FLASHBACK). He defined the participating agents luciferin and luciferase and showed that their reunification in the test tube can produce bioluminescence in the absence of living cells, although he could not know what their molecular nature was. Today we know that luciferases are enzymes which make their substrates, the luciferins, react with oxygen. (Boyle's 'flash' was due to an accumulation of unused luciferin that reacted when the oxygen returned.) That is about the only generalization that can be made about them, as the details of these reactions differ widely between species. The firefly system, for instance, also requires the cellular energy carrier ATP.

Just from superficially looking at the specific properties of the two best-studied systems we can get an impression of this variety. Luciferase from the North American firefly, *Photinus pyralis*, was one of the first enzymes to have its molecular mechanism exposed in some detail, mainly due to research conducted in the laboratory of William McElroy (see FLASHBACK on p. 70) at Johns Hopkins University during the 1940s and 1950s, and later at the University of California at San Diego. The fact that by simply combining solutions of luciferin, ATP, and luciferase in the presence of air which supplies the oxygen, one can get the enzyme to make a product which is easily measured, namely light, made it the ideal object for some of the pioneering work in enzymology. The essential protocol was based on Dubois: just pour the two extracts together and measure the light.

As early as 1960, crystals of the firefly enzyme were available,

Raphael Dubois

I SHOULD ADMIT THAT THE TITLE OF MY BOOK is not entirely new: back in 1914, the French physiologist Raphael Dubois (1849–19??) published a book with the reverse title (*La vie et la lumière*), which also spanned the topics of bioluminescence, photosynthesis, and other interactions between light and living things. While all of these topics must have appeared as massive, mostly unsolved mysteries at the time, Dubois himself was credited with one of the major breakthroughs of the field. The classic, but still too-little-known experiment he carried out in 1885 merits a detailed account.

He dissected a West Indian beetle of the *Pyrophorus* genus and then plunged its still glowing light organ into boiling water—thus extinguishing its light. When he cooled it, he noticed that the luminescence was not recovered. He ground up a second sample at room temperature with a little bit of water and waited for it to lose its luminescence. Finally, he recombined the cold water and hot water extracts in the dark, and saw the beetle's light reappear.

As it was difficult to procure enough insect material for detailed investigation, he then moved on to a marine clam, *Pholas dactylus*, from which he obtained essentially the same result with a slightly different set of temperatures. He called the heat-stable compound luciferin, and the heat-sensitive one luciferase. From his further investigations he concluded rightly that luciferase is an enzyme which catalyses the oxidation of luciferin. In modern terms, his classic experiment would be interpreted as follows: in the hot-water extract, the enzyme must have been denatured irreversibly, such that some of the unused luciferin substrate survived. In the cold-water extract, the enzyme remained active, using up all the luciferin, such that in the end it only contained luciferase. Recombining luciferin with luciferase in the presence of air containing oxygen allowed the reaction to proceed. The only thing Dubois got wrong was that he thought the luciferin to be an 'albumin' (a water-soluble protein like those in blood serum and in milk), while it is in reality a small organic molecule.

While bioluminescence remained a major focus of Dubois's research, and took up nearly half of his 1914 book, he also investigated unrelated physiological issues such as the thermal regulation of hibernating groundhogs. In this field, too, his research (published in a string of papers from 1893 onwards) is still remembered as a fundamental contribution.

which even retained the catalytic activity. Nevertheless, the detailed molecular structure (obtained from X-ray diffraction analysis of such crystals) was only published in 1996, a few months after that of the other unrelated model system, the bacterial luciferase. The structures could not be more different: the firefly enzyme exposed a novel two-part structure reminiscent of a hammer and anvil. Bacterial luciferase, in contrast, consists of two equal parts, each of which has a well-known, barrel-shaped folding pattern. Accordingly, the reaction mechanisms are quite different between the two systems except for the fact that both use oxygen. Bacterial luciferase does not require ATP. It makes the oxygen react with a flavin molecule (its luciferin), an entity that is commonly found as a tightly bound cofactor in enzymes, but rarely as a loose substrate, as in this case.

Sequence comparisons show that the insect luciferase has a great number of cousins with different functions (also making oxygen react with some other molecule, but not producing light in the process), while the bacterial enzyme has so far only been shown to be related to one non-luminescent protein, the function of which remains unknown.

Even after the structures were solved, some questions remained open. Researchers were intrigued by the observation that changing a single amino acid in the sequence could alter the colour of the light that the enzyme produces. Previously held beliefs concerning the interpretation of colour changes as a consequence of subtle shifts between two different forms of the luciferin (substrate), triggered essentially by changes of the local pH, could not account for this observation.

Most puzzlingly, one single species, the click beetle *Pyrophorus plagiophthalamus*, can almost outperform a traffic light using 11 different luciferases (but only one luciferin!), which produce four different colours. In one of the last papers co-authored by William McElroy, Keith Wood and coworkers from San Diego neatly demonstrated this by transferring all 11 genes into separate colonies of the gut bacterium *Escherichia coli*. With identical growth media including luciferin, the bugs indeed produced four different colours: green, yellow-green, yellow, and orange, proving that only the sequence differences between the proteins accounted for their differently coloured light emissions.

Very nice but what's it for?

We tend to produce light for a variety of different reasons and goals. While vision is the one that springs to mind most readily, there are lots of other reasons that may be even more important in certain situations. Just think of all the lights used in cars and in traffic more generally. While the headlights of a car help the driver see in the dark as well as making sure the car is seen, the rear lights clearly don't serve the vision of the person using them, but rather allow them to *be seen* rather than see. Blinkers and reversing lights are used for signalling, as are the flashing alarm signals on police cars and burglar alarms. The red lights in the eponymous districts are guiding the mating behaviour of some individuals, while the illumination of buildings at night mainly deters burglars. Finally, there are some kinds of lighting, like the light effects used in night clubs or as Christmas decorations, which are installed at considerable expense although they don't seem to serve any meaningful purpose.

Similarly, the purposes of bioluminescence vary widely (as do the biochemical details) and sometimes even remain mysterious, so generalizations cannot easily be made. A few obvious 'themes' have emerged from the research, some of which are strangely reminiscent of the human luminescence phenomena quoted above.

- SEEING: Some fish species, like the flashlight fish (*Photoblepharon* and *Anomalops*), use the strong 'headlights' located just underneath their eyes for hunting in the dark. When disturbed, however, they also use a combination of zigzag swimming and light pulses to confuse their enemy. In other fish species, the light is shone to an area clearly outside the visual field of the luminescent individual, ruling out the possibility that it might aid vision.
- SEXUAL SIGNALLING: As I mentioned above, the well-known flashing of the firefly is essentially a mating behaviour. If, for

The Firefly Man—
William D. McElroy

ONE LITTLE BOY must have earned more than $900 in his summer holidays back in 1952. He collected fireflies for William McElroy (1916–99), known at Johns Hopkins University and in the nearby schools as 'the Firefly Man'. McElroy paid 25 cents for 100 of these insects, and that record-breaking boy collected 37 000 of them. Many others did the same on a more modest scale and earned their holiday pocket money this way.

What McElroy and his colleagues did with the fireflies was mainly to take them apart and study the molecular machinery of their light production. Those were the days when even molecular biology started from a whole organism, dissecting it into organs, cells, and organelles, until you arrived at the pure molecular building blocks. McElroy very successfully described insect luciferases and their luciferin substrates, making firefly luciferase one of the best characterized enzymes.

He remained active in this research field through to 1989, in spite of high administrative positions. A graduate from Princeton, he joined Johns Hopkins University in 1946. There he became the founding director of the McCollum-Pratt Institute, founded in 1948 with a major donation from a farmer with the aim of clarifying the role of trace elements in biology. (When recruiting staff for the new institute, he tried to attract one Isaac Asimov, but failed.) The research led to some of the first major health scares of our times. A colleague warned of the impact that the spread of refined flour and sugar would have on public health, and McElroy predicted in 1968 that by 1985 hundreds of thousands of people would starve each year. This must have sounded outrageous at the time, but turned out to be quite realistic.

McElroy chaired the institute and also the biology department after 1956, until US President Richard Nixon appointed him as the director of the National Science Foundation, which he steered through difficult times from 1969 to the beginning of 1972. Then he served as the chancellor of the University of California at San Diego through to 1980, after which, presumably, he was glad to have more time to reignite his passion for bioluminescence.

instance, a male *Photinus* sends out a flash of light, a female may respond, after an obligatory delay time of three seconds, with a longer signal. The male flies in the direction of the signal, stops, sends out another flash, again heads towards the response signal, until …

- ALARM SIGNALS: Fireflies caught in a spider's web will send out a light signal corresponding to a visual alarm cry, like several other species do when caught by a predator. While it is not yet entirely clear where the benefit of this response lies, it could serve to attract a secondary predator which might attack the primary predator and thus free its victim. It appears less likely that the flashing is meant to frighten the predator—they will have got used to it.

- BAIT: Deep-sea anglerfish (again a common name shared by several species, such as *Melanocoetus johnsoni, Ceratias holboelli*, and *Linophryne arborifera*—see Figure 3.1) obviously use the light at the tip of their fishing rod fin to attract their prey. On a more sophisticated level, female fireflies of the *Photuris* species have been reported to mimic the signals of the females of the *Photinus* species, in order to attract their males. For any unfortunate males that are fooled by this mimicry, the sexual signalling will turn into a fatal attraction, as the *Photuris* females will eat them, in order to benefit from a chemical that *Photinus* produces to repel spiders.

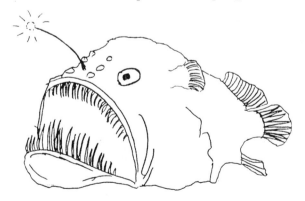

Figure 3.1 The deep-sea anglerfish (*Melanocoetus johnsoni*) uses a lantern to attract prey.

- DECEIVE PREDATORS: Some fish living at a depth where day-light can still be seen as a faint blue glow carry light organs on their underside which mimic the daylight and thus disrupt the dark silhouette which any predator coming from below would use to find and identify its prey. Some squids and crabs, such as the opossum shrimp *Gnathophausia zoea*, squirt their luciferin and luciferase directly into the water, leaving a luminous cloud behind which may hide their flight.

- NO USE: In a number of cases, including, for instance, the lumi-nous fungi, light is emitted without any known benefit. It may be the case that the light is an unwanted byproduct of a reaction which the cell runs for other reasons. The lumines-cent organism may be interested in the chemical resulting from the oxidation of the luciferin, for instance. However, very little is known about these cases.

In summary, bioluminescence is a nice example of the intrinsic randomness of evolution. Chemical reactions that may have had a completely unrelated 'purpose' in the beginning, such as the detoxification of oxygen, may have been recruited in biolumines-cence systems that turned out to be useful and thus evolved to high sophistication. Weighing the advantages and disadvantages of such a trait as light production is bound to involve consideration of complex networks of co-evolution. For each organism that signals using light, there are not only those that are meant to receive the signal, but also those that are not meant to see it.

... and how can we use it?

Fishermen on the shores of the Indian Ocean are known to use the light organ of the flashlight fish, which glows on for several hours after the fish is dead, as a fishing bait. Similarly, Madeira fishermen use pieces of luminescent squid as bait. Japanese soldiers in the Second World War used dried crustaceans of the *Cypridina* genus as lamps—by crushing them between the fingers and then wetting them one can obtain enough light to read by. In all these cases, cells or even light organs of luminescent organisms are simply used up until they run out of fuel, the main benefit being that these systems can produce light with high efficiency and virtually no heat.

Technologically more advanced ways of using bioluminescence tend to reduce it to the molecular essentials and supply as much fuel as needed. Thus firefly luciferase, which depends on ATP and oxygen, can be used in biochemical assays to measure the presence of one of these, when the other is supplied together with the luciferin and luciferase molecules. The ATP assay can be used to check food factories for microbial contamination, for instance.

It has been mentioned above that luciferase was used for some of the first studies of enzyme mechanism, due to the ease of detecting its 'product'. Similarly, it was used in studies of gene expression from 1985 onwards, and later also in related fields such as protein folding and the action of molecular chaperones (folding helpers). Molecular biology companies offer ready-to-use kits which allow scientists to smuggle the luciferase gene into the cells they are studying, and thus use the light emission as a probe.

The main advantages of this approach are that the assay is rapid (results are obtained within minutes), it is much more sensitive than any chemical assays available previously, and it can be used without damaging the living cells or tissues to be investigated. Its limitation is, however, that luciferase luminescence depends on other molecules (luciferin, ATP, oxygen). If any of these are absent, unstable, or not transported to the site of interest under the conditions required for the study, the luminescence experiment may fail. This is why in the mid-1990s another luminescent assay system spread across all molecular biology laboratories within months— a system that is also derived from a luminescent organism, but does not need any 'fuel'.

Green fluorescent protein (GFP) is not a luciferase, as it doesn't create light from chemical energy. It represents the second part of the two-step bioluminescence system of the jellyfish *Aequorea victoria*, converting blue light created by another protein into green light. GFP is a single protein which, after its synthesis in the cell (any cell!), makes a certain stretch of three of its own amino acid residues merge into a very unusual chemical structure which serves as the antenna for the light. Once it has acquired this active structure, it does not require any other molecule to produce its characteristic green light. Just shine blue or invisible UV light on it and it will glow bright green.

In February 1994, Martin Chalfie and others from Columbia University reported that GFP can be used as a marker for gene expression. This means that the GFP gene can be combined with the gene of interest, the pair are smuggled into a host cell, and if the UV lamp (which is available in every molecular biology lab) makes the cells shine green, the experiment has been successful. This demonstrates that GFP is completely independent of the cell type it is in. From bacterial through to human cells, any cell will allow it to fold up to its proper structure, and thus to serve as a 'tag' for the gene or protein of interest.

When researchers realized that this works in any cell or organism, the use of GFP spread like wildfire. Since 1995 it has been available as a commercial molecular biology kit, and the number of published papers that refer to it has grown exponentially. Many molecular biologists immediately preferred this marker system for recombinant genes to the luciferase or chemical assay they used before. But beyond that, developmental biologists jumped at the opportunity to be able to monitor which sets of cells read a certain gene in an embryo, which meant that very soon there were green fluorescent fruitflies, zebrafish, caterpillars, etc., and by 1997 Japanese researchers had come up with green glowing mice which even passed this trait on to their descendants. By contrast, the first transgenic primate, a rhesus monkey born in October 2000 and named ANDi (short and reversed for inserted DNA), failed to shine, even though the GFP gene was incorporated into his genome.

Meanwhile, biophysicists got all excited about the possibility of single-molecule studies using GFP—they found that each molecule goes through rather strange cycles of activity and inactivity. Biophysicists also found out how to modify the colour of the emitted light. By using different-coloured variants of the protein as independent probes, they can now use GFP even to address complex systems where several parameters need to be monitored simultaneously.

Continuing the theme of colour changing, researchers have recently cloned and characterized an equally intriguing cousin of GFP called DsRed. This is a related protein from corals (genus: *Discosoma*), which fluoresces red and is partially responsible for the characteristic pinkish hue of the coral. Detailed studies which the laboratory of Roger Tsien at the University of California at San Diego reported in October 2000 have shown up some advantages and disadvantages of this system in comparison with GFP.

Although fine-tuning GFP can result in a range of colours, none of the GFP variants reach the wavelength range of red light. Hence the 583 nanometres emission of DsRed is a very welcome discovery for anyone who wants to be able to use multiple markers and detect their emissions simultaneously. Exchange of a single amino acid residue can even shift it up to 602 nanometres. Like GFP, DsRed forms its chromophore autocatalytically by fusing neighbouring amino acid into an imidazole ring. It first forms a structure similar to the one in GFP and then extends it to achieve the characteristic long wavelength emission. The latter reaction is extremely slow, which is a drawback for some applications. But positive thinking can turn this into an advantage, as Alexey Terskikh and his coworkers at Stanford University have demonstrated more recently. They deliberately created a slowly changing mutant of a similar red fluorescent protein and showed that it can be useful as a 'timer' in developmental biology. In these experiments, green light reveals cells where the gene of interest has recently been activated and yellow to orange colour shows continuous activity, while red fluorescence indicates the gene has been switched off.

Another potential disadvantage of DsRed is that it likes to associate into tetramers (assemblies of four protein molecules) and maybe even into higher oligomers. Given the amount of experience that researchers have accumulated with the homologuous GFP, however, it is a safe bet to predict that quite soon there will be genetically altered DsRed variants that are monomeric, mature faster, and shine in every imaginable hue of red.

Practical applications of GFP and its younger cousins are numerous and will certainly continue to spread. One of the more surprising benefits of GFP reported recently was the development of a new way of clearing landmines. As all landmines leak small quantities of the explosive TNT, researchers used a *Pseudomonas* strain that naturally feeds on this chemical, equipped it with the GFP gene, and obtained a very efficient biosensor for landmines. Just spread the bacteria onto the mined land (from an airplane). As soon as night falls, a UV lamp will allow you to see the mines glow in the dark, and thus to find and remove them hundreds of times faster than would be possible by conventional methods.

Even more applications are bound to emerge in the near future. Biological applications could involve cell cultures or whole organisms: a recent newspaper item inexplicably listed under 'News of the Weird' potato plants that light up when they want to be watered and a Christmas tree with green glowing needles. Biophysicists could make further use of the unique properties of GFP in single-molecule experiments. The observation that GFP can be switched between different states using light of a specific wavelength, for instance, suggests that it could become useful in molecular signalling or even computation devices in the context of the anticipated development of nanotechnology.

A shining light for education

Apart from technical applications, bioluminescence can also serve as a tool for education. As human beings of all ages tend to be attracted by anything that shines or flickers (witness neon ads and television), several educators have used bioluminescence to develop teaching materials that can treat a range of biological and biochemical subjects.

Fireflies, of course, are a little bit difficult to use, as they only produce their light during the mating season, which coincides with school holidays in many geographical areas. It is also mighty difficult to keep them in captivity, or to manipulate their natural mating behaviour and luminescence. Luminescent fishes are even more difficult to handle and well out of reach for the average school teacher.

Therefore, Mara Hammer and Joseph Andrade have developed

an approach using marine planktonic glow bugs, the dinoflagellates. These protozoa tend to do photosynthesis during the day and produce an aesthetically pleasing blue light at night, implying that they have a biological clock (which we will discuss in the next chapter) to regulate this alternating behaviour. Hammer and Andrade seal these creatures in gas-permeable plastic bags, where they can be kept as a stable ecosystem for a long time, as they require only light and carbon dioxide, which can both enter the bag. Before school visits, they readjust the creatures' clocks by keeping them in artificial light at night and in the dark during daytime, such that in the darkened classroom they can produce an impressive light show.

As they have demonstrated, a wide range of science topics can be treated in connection with the dinoflagellates once the pupils are captured and intrigued by the light phenomenon. Energy efficiency (10 per cent in a light bulb as compared to nearly 100 per cent in bioluminescent organisms whose light produces virtually no excess heat!), closed ecosystems, photosynthesis, and the biochemistry of enzyme reactions (i.e. how do they do it?) are just some of the topics that can be communicated to pupils more efficiently using this approach.

On a more advanced level, of course, molecular elements of bioluminescence can serve to demonstrate procedures of protein expression, purification, and characterization in a visually fascinating way. Both firefly luciferase and GFP have been used for such demonstrations. It turns out that the seemingly eccentric natural phenomenon of bioluminescence may be more useful than many would have thought.

CHAPTER 4

How life is guided by light

SOME TIME IN EARLY MARCH, when I sense instinctively that for the first time in months the sunlight is not only illuminating but also warming my skin, I tend to find a really important scientific paper which I need to read undisturbed by callers, email, or phone—i.e. which I can best read in the University Parks (which I have already enthused about at the beginning of Chapter 2). So off I go with the paper, looking very busy, and head to my favourite bench, where I'll try to find the optimal position that allows me to expose a reasonable amount of skin to the Sun, while giving the casual observer the impression that I am actually reading that very important paper.

When nobody's looking, however, I am just as likely to be contemplating the wonders of how the trees and flowers around me unfold their genetic blueprint by exploding in new sprouts, leaves, and buds. I only mention this, of course, because the process is both powered and synchronized by light. But as plants have a rather obvious interest in light we are not that much surprised to learn that they also can grow and bend towards it, and adjust their clocks by it.

The more intriguing observation is that animals do it too. Like myself and most other human beings, many organisms are seeking the sunlight even though they are not harvesting its energy as plants do in photosynthesis. Why and how they are doing it, and how light more generally influences the actions and the well-being of various living things, will be explored in this chapter.

Like a moth to the fire

Imagine a toy car with two light sensors instead of headlights, and two independent motors each driving one of the rear wheels while the front wheels rotate freely. Now let's wire this up in a way that the more light one of the sensors gets, the faster the wheel located diagonally opposite to this sensor will rotate in the forward direction (Figure 4.1). Place this car anywhere on a table which has a single light source in the middle—what will happen?

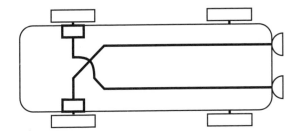

Figure 4.1 A 'cybernetic vehicle' that would show phototaxis. Each of the sensors at the front of the car controls the motor driving the wheel diagonally opposite. If more light makes the motor go faster, the vehicle will turn towards a light source and eventually rush into it.

You don't have to build such a device to figure it out—a bit of mental gymnastics with paper and pencil will show you that the toy car will turn towards the lamp and rush into it like a moth into the fire. This *Gedankenexperiment*, borrowed from Valentino Braitenberg's utterly brilliant book *Vehicles: experiments in synthetic psychology*, demonstrates a simple form of what biologists call 'taxis': an attractive signal emanating from a source makes an organism run towards it. If the attractant is light, we call the phenomenon phototaxis; if it's a chemical, it's chemotaxis, and so on.

Note that taxis requires both orientation and displacement—if a bacterium swimming straight ahead just speeds up when it gets more light, this is a separate phenomenon called photo*kinesis*. If a plant grows or bends towards the light, this is called photo*tropism*, to which I will come back later in this chapter. So, once more for practice: if you run faster in the sunlight than in the shadow, this is kinesis, if you're sitting on a park bench turning your face to the Sun to get a nice tan, this is tropism, and if you turn to the light and run to it trying to get even more light, this is phototaxis. Each of these behaviours comes in a positive (light-seeking) and in a negative (light-avoiding) mode. I, for instance, tend to display positive phototropism and phototaxis, especially in spring after a gloomy British winter, but negative photokinesis (getting lazy in the sunshine).

There are good reasons for both positive and negative photo-taxis, especially if you are living off photosynthesis, and you're mobile but relatively ill equipped to cope with the many orders of magnitude difference in light intensity that nature can throw at you. This is the case for many single-cell algae, and the model organism in which algal phototaxis is most commonly studied is *Chlamydomonas reinhardtii*, a eukaryote commonly found in fresh water and in soil. Its swimming motion was described as early as 1917: the alga uses its two flagella in a breaststroke-like fashion, but swims in helical paths rather than straight ahead. Nevertheless, bio-physicists are still debating the mechanisms which allow the cell to orient towards or away from the light. A recent publication from Rainer Uhl's group in Munich suggests that *Chlamydomonas* can head away from the light or towards it simply by applying more power either to the flagellum closer to the eyespot (the sensor) or to the one pointing away from it. In striking analogy to Braitenberg's 'vehicles', such a mechanism would require a lot less processing power than previous explanations, which required the cell to observe and interpret the sine curve of the light changes as it followed its helical path.

A second microbe whose phototaxis has been studied inten-sively, but has not been fully understood as yet, is the photo-synthetic flagellate *Euglena gracilis*. This unicellular organism is quite unique and can be counted as an alga (as it is a photosynthetic eukaryote) or as a protozoan, as it is mobile and can, in the absence of solar energy, turn to eating other microorganisms. Bluntly speaking, it's a cross between a single-cell plant and a single-cell animal. In darkness, *Euglena* uses gravity to orient itself and swim upwards—knowing that the light it craves is most likely to come from above. In daylight, *Euglena* will use negative phototaxis if the light is too strong for its liking, and positive phototaxis if it's too weak. This enables the microbe to position itself in the Goldilocks zone of the water column at a height where the light intensity is just right for its needs. While the behavioural response is quite straightforward, the molecular details of the photoreceptor and its mechanism have so far remained elusive.

Much as *Euglena* can't make up its mind whether it wants to be a plant or an animal, the next more complex model organism in phototaxis is unsure whether it's a single-cell or a multicellular organism. The cellular slime mould (amoeba) *Dictyostelium* acts like a unicellular microbe for as long as there is sufficient food in the soil where it lives. As soon as some of the cells run out of food, however, they send out a signal which makes some 100 000 of their kind assemble and form a multicellular organism, known as the slug state of the amoeban life cycle. Unlike the single-cell state, the slug is phototactic: it tries to move towards the surface, following the direction from which the most light falls in. Once there, it forms a fruiting body, like a mushroom, containing thousands of cells that act as germs—the amoebal spores. The slug moves towards the surface so the spores can be most readily spread by wind, water, or animals eating them. Some of them will arrive in new areas with sufficient food, where they will multiply like a bacterial culture, starting the life cycle all over again. The phototaxis of amoeba slugs has so far been mainly studied from the genetic point of view. Some genes and proteins essential for the process have been identified, but the full picture still needs to be put together.

The last organism in this rather eclectic group of biologists' pets used in phototaxis research is a water flea living in freshwater ponds and puddles and known as *Daphnia magna*. Its physiology responds very sensitively to environmental stress, and it can therefore be monitored as an indicator species to assess the state of a given biotope. As the flea is quite transparent, its stress response can even be demonstrated in a simple school experiment—you can see its little heart beating under the microscope and count the beats per minute under different environmental conditions. Photo-taxis research with *Daphnia* is mainly addressing the mechanisms underlying the typical vertical migration which many small animals both in marine and in lake environments carry out in a daily rhythm. Interestingly, the swimming movement of *Daphnia* involves a normal oscillation between upwards and downwards swimming. Light changes are believed to influence the overall swimming direction of the flea, mainly by setting this oscillation off balance so that either the up or the down movement dominates.

We will revisit such phototactic movements that depend on the rhythm of daily light changes later on in this chapter, when we come to light-controlled biological rhythms. Meanwhile, we shall leave the world of phototaxis for now—although there would be many more interesting phenomena in other organisms as well—and turn towards tropism.

Turning towards the light

Plants have an obvious interest in knowing where the light is and both growing and bending towards it. These movements are called phototropism. Unlike algae and photosynthetic bacteria, plants have mechanisms for disposing of excess light energy. Therefore they will not normally turn away from strong light, but negative phototropism can be observed in roots, for instance, where it presumably helps to keep the plant upright while the tip bends over to the sunny side.

While photosynthesis mostly relies on red light (plants appear green to the eye because they keep all the red light and reflect the rest of the spectrum!), the phototropic response is triggered specifically by blue light. This was already noted in the nineteenth century by Darwin, among others. The phenomenology of tropic responses has been described in detail and has reached the school books. Nothing easier than putting a few seedlings in a dark room, shining blue, red, or polarized light at them from a well-defined direction, and measuring the lengths and angles in which they grow and turn.

The molecular details of these responses have, however, eluded scientists until very recently. It was only in 1998–99 that a group of blue-light photoreceptors called the cryptochromes (presumably because they have remained hidden for so long) was identified and shown to be important for phototropism in *Arabidopsis thaliana* (thale cress), a plant which is very frequently used for genetic studies, and which in early 2001 became the first green plant to have its entire genome sequence published.

Cryptochromes proved to be an extremely interesting group of molecules for a variety of reasons beyond phototropism. Their evolutionary pedigree links them to a bacterial enzyme, photolyase, that harvests blue and near UV light and uses its energy for DNA repair. Moreover, it was shown that cryptochromes act as a time-piece in biological rhythms not only in *Arabidopsis*, but also in animals such as the fruitfly *Drosophila* and the mouse. We shall come back to this function later in this chapter.

Another family of proteins involved in phototropism was only discovered in the late 1990s. Etiolated seedlings of *Arabidopsis* (i.e. seedlings that have responded to a dark environment by rapid growth with little leaf formation) should normally turn towards the direction from which they are sensing blue light. Scientists observed that some mutants were lacking this response, translated this observation into genetic jargon, and thus ended up naming the genes involved *nph1*, *nph2*, etc., for *n*on*p*hototrophic *h*ypocotyl. What emerged from studies published in 1998–99 was a whole family of proteins (uninspiringly named NPH1, -2, -3, etc.) involved in plant phototropism. The preliminary view of the process is that NPH1 is the primary photoreceptor, NPH3 is an essential protein interacting with it, while NPH2 is involved in signalling further downstream.

The ways in which plants use light for information rather than energy purposes is still under intensive investigation. What has become clear, however, is that you don't have to have eyes to get information carried by light. If plants and bacteria can do this, perhaps animals have light sensors outside their eyes too? (Hang on—is this a shadow falling on the back of my knee?) We will find out later in this chapter.

Not quite an eye

Parents of little mischief makers could definitely use them, motorists would find them helpful too, as would anybody afraid of being stabbed in the back: eyes in the back of the head could come in handy in many situations. Well, if natural selection is going to give us such help, it will certainly take a couple of million years to arrive. Lizards, however, are one step ahead of us, at least in this respect. They do in fact have a photoreceptor at the back of their heads which, although it's not quite an eye, is almost halfway there.

The light-sensitive spot at the back of a lizard's head contains a lens which focuses the incoming light onto photoreceptor cells. We know that this so-called 'parietal eye' cannot allow the lizard to see a visual representation of what's behind it, because the photoreceptor is in fact not connected to those parts of the brain that serve visual perception. (Real eyes and their pathways to perception will be explained in Chapter 5.) It does, however, sense changes of light condition, and responds to them in exactly the opposite way from the response of a normal vertebrate eye (including the pair found on the sides of the very same lizard's head). When a rod cell in a normal eye would respond with a reduced voltage, the parietal eye will increase the voltage, and vice versa.

Therefore, it was generally assumed that the biochemical pathways in the signal transfer from the parietal eye to the brain would be different from those in the visual pathways. Only in 1997 could scientists show that the signalling is in fact surprisingly similar in both systems, involving the key compound cyclic GMP. Similar signalling is also found in the light-sensitive pineal gland of birds.

So some animals have photoreceptors that don't see—but what for? In the case of the lizard, being able to sense the shadow of a predator sneaking up from behind could make all the difference between a premature death and a lucky escape, which provides sufficient benefit in natural selection to account for the existence of the parietal eye. More generally, it is beginning to emerge that the influence of daylight changes on the biological clock of human beings and other animals does not necessarily have to travel via the eye, as will be discussed in the next section. Like the lizard, we may have photoreceptors in funny places without being aware of them. And they may even be very important for our general well-being.

Getting the rhythm

One of the important ways in which light influences us is by setting our inbuilt biological clock, our circadian rhythm. The 'circa' means it only approximately measures out 24 hours, and it needs regular recalibration through the light/dark changes in our environment. As recently as half a century ago, this was a revolutionary new concept, and its importance was established by pioneering researchers including Colin Pittendrigh and Jürgen Aschoff (see FLASHBACK, p. 90). Following these early studies, the circadian control was long believed to be mediated by the visual system, requiring some highly evolved brain response. Surely primitive bacteria which have a generation span of less than a day wouldn't have a circadian rhythm?

Surprisingly, there is a whole big family of bacteria, including several species, that very certainly do know about 24-hour days, even though each cell divides after a much shorter timespan. The cyanobacteria, which we met previously as the likely descendants of those microbes that once invented the two-step photosynthesis mechanism used by today's plants, have a remarkable problem resulting from the combination of their talents. Typically, they can perform both photosynthesis and nitrogen fixation, a combination which practically enables them to live almost anywhere, independent of nutrient supply. The only trouble is that the enzyme needed for nitrogen fixation is easily poisoned by the oxygen produced by photosynthesis.

How do cyanobacteria manage to separate the two processes? From the early 1980s, researchers suspected that they might have a time-switch to do photosynthesis by day and nitrogen fixation at night. When this alternation of tasks was first confirmed experimentally, it was interpreted as a process regulated by the cell division cycle. Only in the late 1980s did a study of the freshwater cyanobacterium *Synechococcus elongatus* reveal that:

- the activity cycle runs on at an approximate 24-hour period even if the bacteria are exposed to constant light and temperature conditions;
- the rhythm can be reset by a different apparent day length;
- the period under constant conditions is independent of the temperature (unlike the rate of cell division).

These are the defining criteria for a true circadian clock. The same could later also be shown for other species of cyanobacteria. So beyond any doubt bacteria, the most primitive lifeform we know, can have an elaborate biological clock.

Just as an aside, it should also be mentioned here that the studies of circadian clocks (not only in bacteria, but also in plants and in animal tissues) have often made use of bioluminescence genes as reporter genes in the way described in Chapter 3. Rhythmic cycles of lighting up in the dark can be used to find out which parts of the genome are expressed under control of the clock, and in which time periods of the clock cycle they are needed.

Key genes of the clockwork mechanism of cyanobacteria have been defined by such methods. Researchers were quite surprised to find that the entire genome of the bacteria is clock-controlled. Some parts of the current overall view of the mechanism still remain largely hypothetical, however, as models are partly based on speculative analogies to the better-known mechanisms in insects.

Along with the fungus, *Neurosporacrassa*, the fruitfly *Drosophila* is the classical object of investigations into biological clocks. The period gene from *Drosophila* was the first clock gene to be cloned (in 1984), and to this day *Drosophila* has the best-understood circadian system. We shall therefore look at it a bit more closely, but only after sorting out the fundamental concepts.

Let's go back to toy cars, and now look at one with a remote control. Imagine you want to keep the car on a long straight road. What do you do? The operating instruction is simple. When you see the car approaching the left side of the road you turn the joystick to the right, and vice versa. Now let's attach a long stick to the end of the car, with a pencil that traces the movement on the road (Figure 4.2). Due to the length of the stick between the rear end of the car

(a)

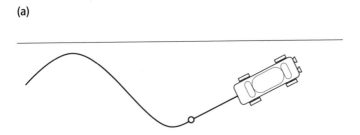

(b)

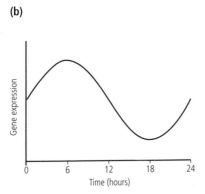

Figure 4.2 Sinewaves generated by negative feedback: (a) toy car analogy; (b) gene expression in a typical circadian clock.

and the pencil, the steering movements made to keep the car on the road will be represented in a slightly exaggerated way, and it will be obvious that they represent a wavy, probably sinusoidal pattern. This is a typical example of negative feedback regulation. That is, if you notice a movement to one side, you steer to the other. (Positive feedback would mean the opposite response and would put your car off the road in no time.)

When driving a real car, you would want to keep the sinewave unnoticeably small so that anybody following you would get the impression that you're driving a straight line (although in physical fact you never ever are, it's always a sinewave!). But when playing with the remote control toy car with the pencil attached, we might take a fancy to the sinewave and want to produce an even bigger one. An easy way to do this is to get drunk so that your response time is increased and you only turn right when the car is already very close to the left border. A slightly more sophisticated way would be to record the car's movement on a video tape which you watch with a programmable delay time of, say, several seconds. With a few seconds' delay the movement would be so obviously sinusoidal that you don't even need the pencil any more to demonstrate this.

So now we've got a nice wavy movement, or oscillation, but we want to control the distance between the repeats of the pattern, so that the car always hits the left border after 10 centimetres, say. I'm not guaranteeing that this would work, but I would try putting a magnet at the left side of the road every 10 centimetres, so it pulled the car over. The location of the magnets should be hidden to the driver, who should still try to steer a straight line.

Now this is exactly the way the clockwork in *Drosophila* works: it is an oscillation produced by a negative feedback loop with an inbuilt delay time, and with a mechanism to adjust the oscillation length (i.e. to set the clock to the actual day length). Just replace the road by a time axis. What oscillate in *Drosophila* are the concentrations of certain proteins and their messenger RNAs. The most important ones are the products of the *period* and *timeless* genes, called Per and Tim. Negative feedback in biological terms means that these proteins need to find a way of inhibiting their own synthesis. And in order to produce a noticeable oscillation of approximately day length, they need to find a very slow way of doing this.

Per and Tim are produced by ribosomes in the cytoplasm, but they do their job in the nucleus of the cell. In order to get there, they must first form a pair (Per–Tim), which then gets imported into the nucleus. There they bind to a pair of proteins called Clock and Cycle, and stop them from doing their job, which would be enabling the expression of the Per and Tim genes. Hence, by acting on Clock and Cycle, Per and Tim are suppressing their own synthesis, thus

providing the negative feedback. The delay time is probably introduced by the fact that Per and Tim only combine slowly to form
Per–Tim pairs, and that a substantial concentration of these pairs is
required in the cytoplasm before they start moving into the
nucleus. The setting of the clock by light happens through light-
induced destruction of Tim. As Per is only stable in complex with
Tim, this protein also disappears when Tim does.

One of the mysteries that remain, however, is the role of the
cryptochromes, which seem to be important for circadian clocks
both in flies and in mammals. They are hot candidates for the role of
the circadian photoreceptors, but they also seem to have a role in
the clock itself, at least in mice, which has not yet been identified.

More surprisingly, it has also remained uncertain for decades
whether insects can sleep like all mammals and birds do. Their
circadian rhythm is clearly important for timing their development
stages, and they also use it to alternate between active and rest
phases. Research published as recently as 2000 now suggests that
an insect's rest is equivalent to a vertebrate's sleep in that it is
controlled not only by the circadian clock but also by an as-yet
elusive mechanism known as the homeostat, which ensures that
the animal gets a fairly constant amount of sleep on long-term
average. How these two mechanisms play together in regulating
our sleep remains to be discovered.

It's all in the hormones

The biggest experiment in mammalian circadian clocks, of course,
has been going on for about half a century with millions of human
guinea pigs, and it's called intercontinental air travel. When you fly
across time zones, your biological clock will stick to the established
day/night rhythm of the place you came from. It will need a few
days (up to one day for each hour of time difference) to adjust to the
different rhythm of the destination. Jet lag and the similar problem
faced by shift workers are blamed for a large number of accidents.

The fundamental principles are the same in mammals, including
humans, as in *Drosophila*: there is an oscillation generated by a

Colin Pittendrigh and Jürgen Aschoff

TIMING IS EVERYTHING, especially when you study chronobiology. So it may be more than a coincidence that the two pioneers of the studies of circadian clocks, Jürgen Aschoff and Colin Pittendrigh, were both born in the 1910s, died in the 1990s, and managed to find a science-related activity important enough to keep them out of the Second World War. For each of them this wartime activity led to a life-long research interest in circadian clocks.

Jürgen Aschoff (1913–98) studied the physiology of heat regulation in humans during the war and was ruled indispensable for the training of medical doctors. His discovery of 24-hour regularities in heat loss led him to study biological timing in human volunteers, including himself, demonstrating that our biological clock will continue to run in an approximate 24-hour rhythm even in the complete absence of time cues. Aschoff's department at the Max Planck Institute for Behavioural Physiology, which Konrad Lorenz founded in the early 1960s, became a major centre for chronobiology.

Colin Pittendrigh (1918–96) was sent to Trinidad as a botanist, and spent a year investigating some peculiarities in the reproduction of banana hybrids. He then went on to study mosquitoes that spread malaria. In both studies he noted biological activities linked to a specific time of the day. In the mosquitoes the timing was even specifically different for each of the species studied. Similarly, in his Ph.D. work with Theodosius Dobzhansky after the war, he noted that two different species of the fruitfly *Drosophila* have their activity peaks at different times of the day. During the 1950s he could establish the concept of a circadian clock as we know it today, involving internal oscillators (that run on in the absence of time cues), light-induced resetting, and temperature independence.

Seeking to generalize the new concept and to establish it against those who thought a biological clock was an unnecessary complexity, his lab also found such mechanisms in the microbe *Euglena*, and in the bread mould *Neurospora*, which was to become a major model system for the genetic studies of biological clocks.

Pittendrigh's main concern, however, was evolution: what is the benefit of having an inbuilt clock? Why do nearly all higher organisms have it? One of the issues very hotly debated in the early days was the sun-compass orientation in birds and insects.

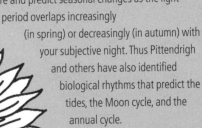

The Sun is no use for orientation unless you know the time. Thus Pittendrigh quotes a lecture by Gustav Kramer on birds which use this method as a major inspiration in his search for 'the clock'. This quite clearly defined purpose helped to win over those who initially failed to see the point of having an internal timekeeper. Other opponents still attributed the animals' sense of time to external factors, such as diurnal changes in the Earth's magnetic field, or an unknown 'factor X'. Only when molecular genetics began to reveal the individual cogs and wheels of the circadian clock could these objections be ruled out completely.

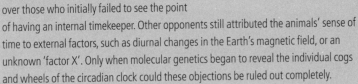

Another competitive advantage of having a clock is that you are able to anticipate when repetitive changes will happen that may affect your life. While you could follow the daily cycle just by monitoring the height of the Sun, an inbuilt clock also allows you to measure and predict seasonal changes as the light period overlaps increasingly (in spring) or decreasingly (in autumn) with your subjective night. Thus Pittendrigh and others have also identified biological rhythms that predict the tides, the Moon cycle, and the annual cycle.

negative feedback loop with a time delay, and calibrated to day length by light signals. Research emphasis has been somewhat different from the clock research done on flies. Especially for the human system, the questions asked are not so much about which gene product cycles in which phase, but are rather more practically oriented, that is, referring to the ways in which the clock output signal affects our lives. This is particularly relevant to problems such as jet lag, shift work, and 'seasonal affective disorder'. The most important output signal of the human circadian system is the hormone melatonin, which was hailed almost as a wonder drug for such problems in the mid-1990s.

One key element of the biological clock in humans and other mammals is a pea-sized structure right in the middle of our head, between the cortex, brain stem, and cerebellum: the pineal gland. In evolutionary terms it is related to the lizard's parietal eye. In chickens it has indeed retained some sensitivity to light, but in humans it is buried so deep under all those masses of cortex we are so proud of that there is no chance of it detecting any light even if it had the ability.

Interest in the pineal gland is as old as science itself. René Descartes (1596–1650) proposed (correctly, as was confirmed three centuries later) that it may be connected to the eyes. He also described it as the seat of the soul, a notion which we find harder to accept, unless you believe that alarm clocks have souls too. (Considering that Descartes believed in a clockwork world (probably for the same reasons why we tend to use computer metaphors to grasp the brain and other complex systems—clockwork happened to be the high-tech of his time!), it is quite ironic that he took such an interest in a structure which was only identified as part of a clock more than 300 years after his death.) It was a rather straightforward argument: you only have one soul (he didn't know about multiple personality disorder), but most of the substructures of the brain come in symmetrical pairs. Being a singleton is what made the pineal special to Descartes.

In 1958, Aaron Lerner and his coworkers at Yale succeeded in preparing a few milligrams of a novel hormone from the pineal glands of some 250 000 cows. The active principle the researchers had been chasing was very potent at discolouring frog skin cells known as melanophores; hence they called it melatonin. The substance turned out to be a small-molecule hormone present in all vertebrates, and interesting for many different reasons, most of them more important than lightening frog skin.

The pineal gland converts serotonin (a neurotransmitter linked to happiness and, er, to the time lag between male orgasms) to melatonin and secretes the hormone at night. It doesn't know day from night by itself though. It gets the rhythm from another brain structure known as SCN (short for 'suprachiasmatic nucleus', because it's located just below the 'chiasma opticum' where some of the optic nerves cross over), which appears to be the main body clock in mammals and gets its time cues from the eyes.

Melatonin can reset the circadian rhythm, so one way of treating time-shift problems, which was pioneered by Josephine Arendt's group at the University of Surrey, is to swallow a small dose of the hormone at bedtime. This has two main effects: it acts as a sleeping pill, making you fall asleep quite quickly, and it induces the pineal gland to produce even more melatonin, which effectively means the body clock is reset. Typically, 5 milligram fast-release doses of the hormone are taken at bedtime for four days after a flight. The hormone itself, being a natural substance, is not patentable. Therefore drug companies have been reluctant to invest in proceedings to have it officially approved as a drug. It is available from healthfood shops in the USA, however, and is quite commonly used there.

Meanwhile, Arendt and her group are collaborating with industry to develop a more efficient (and patentable) drug based on the hormone. While she is enthusiastic about the benefits that a carefully designed melatonin treatment can have for 'free running' (i.e. not naturally synchronized to the 24-hour day length) blind people, shift workers, and other people with clock-related sleep problems, she insists that the benefit depends on knowing the precise biorhythm status of the patient, as a dose given at the wrong time can mess things up even more. Ideally, patients should be able to monitor their circadian rhythm, and then apply melatonin (and avoid bright lights) at the appropriate times. And, by the way, melatonin will neither stop you from ageing nor prevent cancer. The various 'melatonin miracles' which created some amount of hype during the mid-1990s have so far failed to be supported by experimental evidence.

Apart from resetting your clock and making you feel sleepy, melatonin also suppresses nerve cell activity in the SCN, thus presumably preventing a flash of lightning at midnight advancing your body clock to dawn. The details of the mechanisms by which the clock responds to light and melatonin, however, are far from being fully understood. For instance, the light receptor responsible for resetting the clock has remained elusive, as mouse studies suggest that neither rods nor cones of the retina are required for the resetting of the SCN. More confusingly, one study published in early 1998 suggested that shining light onto the back of a person's knees (while the person does not know whether the light is on or off) can very efficiently reset their circadian clock. If confirmed, this would make the hunt for the elusive light receptors even more intriguing, while offering potential benefits in the treatment of jet lag and other biorhythm disorders.

Keeping the 24-hour rhythm is not the only function of the circadian clock. It also allows us to 'feel' the seasonal changes, as the day length changes in comparison with the circadian rhythm. Thus a better understanding of the ways in which light interacts with our hormones would also help to beat the 'winter blues', otherwise known as seasonal affective disorder (SAD). This condition affects between 2 and 10 per cent of people living in northern Europe, with sleep problems, overeating, depression, lethargy, and behavioural/ social problems as the most common symptoms. It is generally believed to result from the absence of bright light, which is why relatively gloomy places can still be OK to live in if they have lots of snow in winter. Apart from the home remedies of eating chocolate or booking holidays in sunnier places, SAD can be treated by illumination with especially bright lights available from companies such as 'Inside Out' in Cambridge, UK. Although medical use of light therapy for a number of conditions has been pioneered by the Nobel laureate Niels Finsen more than a century ago (see FLASH-BACK), some treatments remain rather poorly understood and non-specific. If one day we understand the workings of our body clocks

Niels Ryberg Finsen (1860–1904)

THE NOBEL PRIZE IN PHYSIOLOGY OR MEDICINE for the year 1903 was attributed to Niels Finsen, who was already too ill to attend the ceremony and died less than a year later. It was, however, his ill health which set him on the path to the highest scientific honour, which he received for pioneering studies of the curative effects of light.

Born in the Faroe Islands to an Icelandic family, Finsen was educated in the Faroes, Iceland, and Denmark, and took up medical studies at the University of Copenhagen in 1882. From 1890 to 1893 he held a university post as a prosector in anatomy, but then abandoned it to devote more time to research, making a modest living from private tutoring.

From 1883 onwards, if not earlier, he suffered from Pick's disease, a thickening of connective tissue which interferes with the function of the liver, heart, and other organs. His first symptoms included anaemia and general tiredness. As his accommodation was north-facing, he began to think the symptoms could be alleviated if he received more light. A search of the available physiological literature, however, taught him that no such link was known to science. From around 1888 onwards he systematically collected and analysed all the information he could get on light-seeking behaviour in animals.

Although there was no theoretical foundation in physiology that would have predicted any effect of light on human health, he finally moved into clinical studies and was able to demonstrate beneficial effects of UV light in the treatment of smallpox and lupus vulgaris (a tuberculosis of the skin). In both cases, excessive scarring can be prevented by light therapy, partially because of the bactericidal effect of UV light.

By 1896 he had founded his own research institute, the Finsen Institute, in Copenhagen, which continued these studies on a larger scale, benefiting from both government funding and private donations. Today, phototherapy has a number of applications ranging from neonatal jaundice to tuberculosis.

better, people suffering from the gloomy winters might be cured by carrying a small device that would shine the right doses of light in the right place at the right time of their circadian rhythm.

Getting the right dose

As with many substances, the benefits and risks of light are strongly dependent on the dose you get. Too little light can make you ill, but so can too much. Evolution found a simple way of getting it right: adjusting the skin colour. But then we messed it all up, by sending fair-skinned people to live in Australia or to get sunburnt in Ibiza. Knowing how much light is good for each of us should help us stay healthier for longer.

Too little light, apart from giving you the winter blues, can also endanger your physical health. We need vitamin D to be able to take up calcium for the bones and teeth. The chain of reactions by which the body produces this vitamin begins in the skin with a derivative of cholesterol—known as a swearword to the health-obsessed, but essentially representing a chunky wedge-shaped molecule of four merged rings to the chemist. One of these rings can be opened with the help of UV light, resulting in a molecule known as calciferol, which doesn't look or behave like cholesterol at all. After some subtle, enzyme-catalysed modifications round the edges, we arrive at vitamin D. Thus, it is clear why lack of UV light can lead to vitamin D deficiency symptoms.

Light deprivation of pregnant women may even harm their babies. Some researchers think that the as-yet unexplained over-representation of spring birthdays among patients with schizophrenia could be an effect of the lack of UV light their mothers experienced during pregnancy. Next to living in a big city, a spring birthday is the biggest non-genetic risk factor for schizophrenia.

At the other extreme, having too much UV light, especially of the higher-energy UV-B variation (wavelength range 280–315 nanometres) is a well-known trigger of skin cancer. Epidemiological research on fair-skinned Australians—who have the highest risk of skin cancer in the world—demonstrated early on that sunlight was involved, and that exposure during childhood had something to do with the cancer developing decades later. British people who emigrated to Australia after their eighteenth birthday carry a risk very similar to those who stayed in Britain, while those who moved in early childhood have the higher risk typical of Australian-born people with fair skin. What the light does to the children's skin on a cellular and molecular scale was a lot harder to figure out and is not yet entirely clear.

The DNA of our cells absorbs UV light quite efficiently and can undergo chemical reactions using its energy. A quite frequent light-induced reaction merges two neighbouring cytosine (C) bases into a pair of conjoined twins, a structure called a pyrimidine dimer. In the next copying step, this dimer will be misread as a pair of thymine (T) bases and matched with a pair of adenines (A) on the opposite DNA strand. Hence, in all subsequent copies, the TT sequence will replace the original CC sequence.

Most DNA damage can either be repaired (an enzyme called photolyase can find pyrimidine dimers and split them, again using the energy of light) or trigger the cell to commit suicide (a process known as apoptosis). However, the repair system itself has an error rate too, so some damage may persist. And if the damage knocks out a protein involved in the apoptosis pathway, things can turn ugly.

For some kinds of skin cancer the signalling protein $p53$ has been identified as the target of the damaging UV rays. This is an extremely important element in the signalling pathways that lead from the detection of DNA damage to the appropriate response (repair or apoptosis), as witnessed by the fact that $p53$ is found mutated in more than half of all human tumours that require medical attention.

In non-melanoma skin cancers, there appears to be a two-step mechanism that leads to malignancy. In the first step, a CC pair in the $p53$ gene of one skin cell is knocked out and replaced by TT in its descendants. The cell will still behave normally under normal circumstances, but it has turned into a time bomb, because it has lost the ability to commit suicide when its DNA is seriously damaged. Later on in life, the cell will suffer more damage from sunlight.

Other genes may be disrupted. Neighbouring cells with similar damage will die from apoptosis, but the cell with the disrupted *p53* gene, being unaware of its damaged state, will multiply and fill the space left by the suicides of the others. At some point, in one of these mutant descendants, a further mutation will open the path to uncontrolled, cancerous growth. Typically, this can happen several decades after the initial mutation of the *p53* gene.

This is just one of the ways in which UV light can lead to skin cancer. There are certainly other genes that can be involved, and it has to be said that the molecular mechanisms of the deadliest kind, the malign melanoma, are still unclear. More research needs to be done.

To a certain extent, a dark tan can protect from skin cancer—this is why the Aborigines have much lower risks than the Australians of northern European descent. However, as skin cancer only develops late in life, at an age which early humans rarely reached and which certainly was way beyond their reproductive age, this does not provide a satisfying evolutionary explanation of why people have different skin colours. Even though this is an obvious question to ask (and many children, including me many years ago, have embarrassed their parents by asking it in the wrong place at the wrong time), scientists have only very recently found consistent answers.

If you draw a map of the world showing the intensity of UV radiation received on average in different places, this will broadly agree with a map of how dark the skin of the native population is. This tells us, first of all, that UV is probably involved—but how?

Having a dark skin and very little UV light might get people into trouble with their vitamin D synthesis, as explained above. This might explain why people get lighter away from the tropics. There is a very instructive exception found in the Inuit of Canada and Greenland. They are too dark for the place they live in and shouldn't be able to survive there. It turns out that they are getting most of the vitamin D from their diet of seafood—without which nobody would be able to survive at the latitude where they live. Thus the vitamin D synthesis provides a consistent explanation of why skin gets whiter.

But why does it get darker or stay dark, considering that there is no danger of getting too much vitamin D via the skin route? Skin cancer is an unlikely answer. When our ancestors spread out over the globe, they were much too busy making a living and colonizing new continents to consider sunbathing as a major pastime the way we do. And even if they got skin cancer in their fifties, that wouldn't have cut that many years off their lifespan, nor affected the fitness of their children that much.

It was only about a year ago that researchers found a more plausible explanation. UV radiation can destroy folic acid, another vitamin from the family of B vitamins. As anybody who has had children knows, folic acid is quite important during pregnancy, as a lack of it can lead to lethal malformations of the baby's nervous system. As these malformations are less frequently observed in dark-skinned people, and as they have an obvious impact on evolutionary fitness, they could quite possibly be the main driving force towards dark skin.

Still, the matter is far from resolved. Other researchers have proposed alternative explanations. One, for instance, suggested that the cellular compartments carrying the skin pigment, the melanosomes, might also have a role in fighting pathogens. This appears to be confirmed by the observation that dark-skinned people are less likely to get skin disease.

Whatever the ultimate answer, genetic studies have quite overwhelmingly shown that skin colour can adapt relatively quickly in evolutionary timescales, so it is absolutely no use in defining kinship between populations. Cavalli-Sforza and other geneticists have shown that the genetic variability between native African ethnic groups is in fact much greater than in the entire rest of the world. Thus, the concepts of a 'black race' and a 'white race' are scientifically about as meaningful as classifying people into those who wear sunglasses and those who don't.

If you can't stand the light

On 5 July 2001 Hannelore Kohl, wife of the German ex-chancellor who had been in the top job for more than 15 years (1982 to 1998), committed suicide. As the media were quick to point out, she had lived in the shadow in more than one sense. Firstly, for decades she had eclipsed her own intelligence and put her career aside to serve as a model housewife to the steamroller of ambition that was her husband. But for the last year of her life she had also lived in the shadow in a more literal sense, as she had developed a serious allergic response to daylight after being given a course of penicillin. She was forced to live in mostly darkened rooms, going out only at night, and communicating with the rest of the world mainly by telephone.

No details of her disease have been made public, and some experts have cast doubt on whether it was properly diagnosed and treated. But it has drawn public attention to the previously little-known fact that for a small number of people even small doses of sunlight can cause suffering and death. While the allergic reaction to light has remained mysterious, there are other cases where the problem is more clearly defined.

The most prominent and best-studied example of a fatal light-mediated disorder is the rare genetic disease known as Xeroderma pigmentosa, which affects the DNA repair pathways and essentially leaves sufferers completely vulnerable to UV light. Unless diagnosed very early on, they are most likely to die from skin cancer during childhood. If the gene defect is known from the beginning, however, protective clothing can enable them to lead as normal a life as you can have in a spacesuit. It is estimated that there are only around 2000 sufferers worldwide. The reason why it is so well described lies in the simple fact that it is a cancer with a clear genetic origin, offering a rare opportunity for early research into the molecular biology of cancer. Some more common but less serious diseases made worse by sunlight include lupus and polymorphous light eruption.

While some of the Sun-related diseases are relatively well understood, a few others have remained as enigmatic as the suffering of Mrs Kohl. Considering the importance that sunlight has for our survival and well-being, it is rather shocking that we only have a fragmentary understanding of how it interacts with our physiology. Clearly, there is a lot left for researchers to elucidate.

* * *

POSTSCRIPT: Oh, and in case you wondered: I don't spend all summer on that bench. By June I've normally had enough sunlight to refuel, and keeping the tank full will be possible just with the occasional rays I catch cycling to and from the office or walking with the children.

CHAPTER 5

Seeing and perceiving

COULD YOU DO ME A FAVOUR and imagine a pink elephant for just one moment? Yes, that's right, the entire animal coloured in the most horrid pink you can dream of. And make him wear an oversized black bowler hat. Done? Thanks. The amazing thing about this is, of course, that I can create a visual image in your brain, even over long distances in time and space, and even of things which neither you nor I have ever seen.

You are able to read the words on this page (like 'pink elephant') because light is reflected from them, and focused by your eyes onto your retina, where it creates a miniaturized upside-down image of the letter shapes (no pink elephants involved so far, just black dots on a white background). The presence of this impression on the retina is converted to nerve signals, which, over various processing steps, will arrive in a region of your brain which is capable of dreaming up a pink elephant by combining visual elements of real (grey) elephants you have seen with your experience of colour impressions (and historical headgear).

As we have seen in the previous chapters, many organisms respond to light in a variety of ways, but only some can make use of the information content which light carries when it is reflected, refracted, or absorbed by objects in its path. To make use of this information, you need eyes (at least one), and also a fairly sophisticated brain, and these standards are only achieved by animals, and not even by all of them. Within the animal kingdom, however, there are many different kinds of eyes and visually competent brains. While their workings are not yet understood in full, many interesting aspects of biology and psychology are touched by the investigations into this field.

For anybody who remembers fundamental optics from school physics, the process of seeing may appear deceptively simple. The eye contains a lens, which projects an upside-down image of the world onto the retina, much like the lens of a conventional camera projects an image onto the film. This is, however, not nearly the whole story. We shall find out in this chapter why a human head is not like a video camera or, in other words, why seeing and perceiving is far from being the same. In the pink elephant experiment we did above, for instance, the image created in your mind was very different from the one on your retina. We shall start by zooming in onto a property of light that has not played a significant role so far: its information content.

A brighter way of using light

Some 200 years ago, the French priest Claude Chappe (1763–1805), having lost his job for revolutionary reasons, turned his interest to physics. He investigated the transmission of current through wires, but appeared to be less than satisfied with this technology and started looking for alternatives. In March 1791 he demonstrated the first optical telegraph: a signal board, which was to be turned from black to white after certain time intervals, thus coding for the numbers from 0 to 9.

After some arguments, he was able to convince the National Convention of the first French republic that his idea was useful for the rapid transmission of military orders. Thus he received funding for the first telegraph line with three stations. This line, using an improved signalling system no longer depending on time measurements, started operating in July 1793. The new technique, for which Chappe introduced the word *télégraphe*, allowed the operators to transmit up to 92 different signs in an uncomplicated way (Figure 5.1).

When Napoleon came to power, he invested massively in the new technology. Thus, by the middle of the nineteenth century,

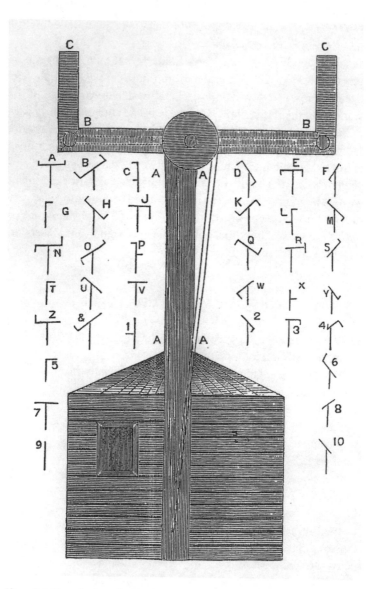

Figure 5.1 Chappe's telegraph and coding system for letters. Allowing only vertical or horizontal positions for the central bar, and 45 deg. changes for the two arms, Chappe defined 92 easily distinguishable signals. Apart from the letter code shown here, there were also code pairs for frequent words or sentences, allowing a data transmission speed similar to that of electric telegraphs.

France was equipped with some 4800 km of telegraph network linking 26 cities with Paris. It had 556 signalling stations, typically small, stone-built towers of about 5 m high, on top of which the characteristic signalling arms rose to an additional height of a couple of metres. Some of them can still be seen—the one I have fond memories of serves as a youth hostel in Les Sables d'Olonne, on the Atlantic coast. A similar, but smaller network was built in Sweden, along the Baltic coast. Replacement of these networks by electric cable telegraph only began in 1846.

For us, living in the age of the internet, the arm-waving technology appears somewhat Neolithic at first glance. On the other hand, we are well prepared to recognize the revolutionary power which the possibility of transmitting messages across distances of hundreds of kilometres and within minutes must have had at the beginning of the nineteenth century, when a galloping horse represented the upper speed limit on everything. An information revolution *avant la lettre*, so to speak. Much as in our present-day information revolution, where the internet got the better of the earlier French 'minitel' system, the French invention was the first on the market and then got swept away by the American competitor.

And it is particularly remarkable that this very first highly efficient telecommunications technology did not use electricity, which had just become available and might have triggered such a revolution, but two elements that we shall discuss in the rest of this chapter—namely light and the human eye. The idea was so obvious in principle that one wonders why neither the Romans nor the Renaissance scientists had thought of it before. After all, seeing is—next to language—the most important means of information transmission for most of us.

So how can light carry information? The light that the Sun shines down on us is a broad mixture of very different wavelengths and orientations, emitted with nearly constant intensity. Not much information in there for our eyes (except if you analyse the wavelength distribution in detail to find out what kinds of reactions within the Sun have produced them—see Chapter 1). When this jumble of various electromagnetic waves that we normally call white light hits an object, however, certain parts of it may be swallowed up (absorbed), sent in a slightly different direction (refracted), or thrown back (reflected). Therefore, the light that reaches our eyes after interacting with the object in question carries information about the colour, transparency, refractive properties, shape, etc., of the object. All these properties we can interpret quite efficiently with our eyes and brains. There are, however, other levels of information content, such as the planes of polarization, which remain hidden to us, but can be used by bees.

One advantage of light as an information carrier is hardly important in nature, but more and more in technology: its speed. Since Einstein's relativity theory we know that the speed of light in a vacuum is the highest speed that anything can move, so information carried by light is literally carried as fast as possible. Ironically, the optical telegraphs were faster (at least in the transmission from one station to the next) than the electrical ones which replaced them.

All these advantages of light as a medium for information are good reasons for evolution to develop some means of using it. Especially for animals, which are typically busy eating other organisms and while avoiding being eaten themselves, the optical information about their environment is of paramount importance. Thus it is not too surprising that evolution has developed eyes several times independently in different lineages of the animal kingdom.

Various ways of building an eye

The eye has always been a favourite example in the battle between evolutionary biology and creationism. I don't really feel the need to roll up this debate here, especially because two buildings down the road Richard Dawkins spends much of his time writing books to drive home the point that yes, a complex organ like an eye can be explained elegantly by the theory of evolution.

And yet creationists still tend to say: 'What use is half an eye?' All I want to point out here is that of course there are lots of half and even quarters or smaller fractions of what we regard as a fully formed eye around in nature. There are light-sensitive organs which are more simply built than our eyes in that they don't have a focusable lens (snails and mussels), can't distinguish colours (cats), or even can't see anything except changes between dark and light (the third eye which lizards have at the back of their heads, which we discussed in Chapter 4). And of course even a primitive light/dark distinguishing sensor (not even the hundredth of an eye) would have given any primitive ancestor of today's vertebrates an important advantage. The shadow of a predator could enable the animal carrying such a sensor to flee in time, while the one without it would be more likely to be caught.

From the countless different kinds of light receptors found in nature, let us just consider three examples representative of larger groups here:

- the human eye as a representative of the 'camera eye' equipped with a lens adjustable for distance and an aperture;
- the eye pits of the invertebrates, which in some species have evolved to the degree of including a lens;
- the compound eyes of insects, e.g. the bee.

We shall see that all these have their advantages and disadvantages. Before we are tempted to regard the human eye as an optimal construction and the measure of all things, we should remember that we cannot see as well at night as cats, that we are blind to polarization differences of light, and that our retina is arranged in a way that in engineering terms would clearly be called the wrong way round, while certain octopus species have similarly highly sophisticated camera eyes without this defect.

A light beam that meets a human eye will first pass through the cornea and then via the opening in the pupil into the lens (Figure 5.2). Naïvely following the camera analogy one would tend to assume that the lens accounts for the refraction of the light which allows the image to be projected onto the retina. This, however, is not the case, as the strongest refractive effect occurs on transition from the air into the cornea. The refractive index, which dictates how strongly a beam of light is bent when transferring from one

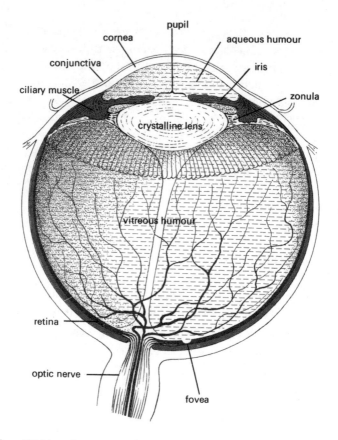

Figure 5.2 Schematic cross-section of a human eye.

medium into another, is roughly proportional to the density of the medium. It increases in the series vacuum, air, water, glass, for instance. The difference in diffractive index between air and cornea is a lot bigger than between the inner components of the eye, which are all essentially water-based solutions. In other words, a camera lens has air on both sides and therefore does a lot of refracting, while our eye lens is surrounded by materials of similar density (the cornea on one side and the vitreous body on the other), and therefore can only refract a little bit.

The main function of the lens in a human eye is to accommodate to different distances. In cameras and certain fish species, accommodation is achieved by moving the lens back and forth. In our eyes, however, the lens stays in place and only changes its shape. A ring-shaped muscle, the so-called ciliary muscle, adjusts its shape in such a way that the image on the retina is in focus. Neither the cornea nor the lens is connected to the blood stream. The lens contains dead and dying cells in its middle, and refracts mainly by virtue of its enormously high protein content (see next section).

Between the cornea and the lens we find the iris, which like a camera aperture regulates the amount of light permitted to enter the eye. The pigments in the iris determine the eye colour. The pupil, of course, is no structure at all, but only the circular hole in the middle of the iris. As we are able to look 'out of' this hole, it is intriguing that we cannot easily look into it, i.e. that we cannot see somebody's retina by looking into their eyes and see only a black circle instead.

For this you can blame either the structure of the eye in question or the head of the observer. As the eye that we look into focuses light like a magnifying glass, we could from any direction only observe a small area of the retina. The light to illuminate this area so that it becomes visible would need to come from exactly the same place as the observing eye. One might say, therefore, that the black of the pupil is the shadow that the observer's head casts onto the particular area of the retina that s/he's looking at. Ophthalmologists and opticians can of course bypass this problem by peering through an ophthalmoscope (invented in the mid-nineteenth century independently by Charles Babbage and Hermann von Helmholtz: see FLASHBACK, p. 130), which is essentially a semi-transparent mirror allowing them to shine light into the eye from the direction they are looking from. Thus they can, for instance, check the retina for the onset of any detachment.

What they see (and the patient also sees in stray reflections) is a mesh of small blood vessels, a network that has given the retina its name (Latin *rete*, for 'net'). At higher magnification, they would also see the nerve endings and the back of the light-sensitive cells (rods and cones). Much like a modernist 1960s building with all the pipework and structural pillars deliberately laid bare, all the wiring of our retina is on the outside, which in this case is an example of an engineering blunder of evolution, because these structures are in the way of the light that needs to get to the pigments that are on the far side of the retina. What's worse, there is one place in each retina where all these nerves have to pass through it so there cannot be any light sensors. This is called the blind spot, and it is located near to the middle of the eye on the inner (nose) side. This intriguing design is found in most vertebrate eyes, and can be explained in terms of embryonic development and the fact that vertebrate eyes have evolved as part of the brain, unlike other kinds of eyes, which evolved from specialized skin cells, to which we shall turn now.

Before we had cameras with lenses, there was the camera obscura, which was already used by Renaissance painters. This is essentially a narrow hole which projects an upside-down image of any sufficiently bright scene into any sufficiently dark room.

Nature has also been using camera obscura-type eyes for a long time, but even before that she invented the biological counterpart of the film. The most primitive eyes can be seen as gradual improvements springing from simple light-sensitive skin cells. Obviously, if you only have a flat surface of cells that can respond to light, it's not an image-forming eye, but only a sensor for brightness. But start converting this patch of sensitive cells into a concave bowl shape, and you begin adding a sense of direction. Light from the right-hand side will be more strongly detected on the left side of the bowl, and vice versa. Such pit-shaped eyes are very common in invertebrates such as worms.

Make the bowl grow into a full sphere, and you have a living implementation of a camera obscura. Allow the skin to grow over the hole, and for a lens-like thickening, and fill the cavity with a transparent secretion, and you have an eye in the style of a cheap fixed-focus camera. Thus far the development of eyes has occurred many times in parallel. It was a rather obvious thing to do, once you had invented light-sensitive cells, so it happened in various branches of life's family tree in very similar ways.

A fundamental split occurred, however, over the question of how to proceed from there towards the highly sophisticated eyes of today's most successful visual species. The vertebrate approach which I have outlined above, and to which I will add more details below, essentially works on the basis of adding bits and pieces to the camera obscura eye. The most important extras included in our eyes are the adjustable lens, the shutter, and the specific sensors for three different colours.

Insects, on the other hand, have taken the primitive eye and multiplied by more than a thousand, resulting in the so-called compound eye. Each individual component sensor (ommatidium) is in charge of one directional segment, as in a major astronomical observatory, where each telescope scans only a small section of the sky. They are insulated from each other by pigmented layers, and each has its own set of light sensor cells. This arrangement, especially as it is normally contained in a very small space, provides less resolution of fine detail than the average mammalian camera eye (which, however, is also thousands of times bigger in volume terms).

In some ways, however, insect eyes can outperform ours. One is temporal resolution. Fast-flying insects can take 300 snaps each second, while even the fastest mammals can't do better than 50 per second. Humans take in about 10 pictures each second, which is why the sequence of images in a movie, changing over at a rate of 24 per second, appears to us as a continuous movement. The title of a German science book I came across recently translates roughly to: *Why flies get bored at the cinema*. From the above figures the answer is obvious: for them the screen shows the same picture for ages, then blackness for a good while, then another picture. They wouldn't see a movie at all. So better leave your flying pets at home when you go there.

The other area where at least some insects beat us is polarization. If you pass light through a polarizer, only light waves oriented in a certain plane will get through, and such light is called (linearly) polarized light. To a certain extent our atmosphere has this effect too, which is why the analysis of light polarization from a segment of the sky can give you information about the height of the Sun even if it is behind clouds. Bees can use this information contained in light polarization in combination with their sun-compass. That is, they can combine the time (given by their circadian clock— see Chapter 4) and the position of the Sun to determine and follow a precise compass direction. Polarization effects are strongest in the

UV range, and it is thus not surprising that bees can see in this range and use it for navigation.

Fascinating as bees may be, I'll leave them following their compass wherever it takes them and will now return to the vertebrate eyes to look at some of its components in more detail.

A lens is for life*

Now let's follow the path of the light meeting the eye through to the conscious vision in a more leisurely way again, so we can look at some of the touristic sights along the way. First, having crossed the cornea, we will consider the eye lens, which, as I explained above, contributes very little to the overall refractive force of the eye, but is absolutely crucial to our ability to adapt to viewing objects at different distances.

The eye lens is a rather intriguing tissue. You probably know that cells are rubbed off your skin all the time, to be replaced by new cells growing from the deeper layers of the skin. Similarly, most tissues of the body have a certain turnover rate, replacing dying cells with fresh ones. Not so the eye lens. It keeps growing at a very slow pace from the inside out, but it never replaces any of the old cells, or even their proteins. Even the very first lens cells that form during fetal development will stay intact for a lifetime, possibly for a century. Their metabolism will be all but closed down, so you could call them dead cells, but they still manage to keep their transparency, which is their most important function.

This poses some interesting questions for biochemists to scratch their heads over. One could dream up a way of keeping a cell alive for a century if the cell was allowed to replace its molecular components from time to time, as most cells do. But lens cells don't even do that. Once they are mature (and engulfed by younger cells nearer the surface of the lens), they have turned into a bag of proteins, and these proteins will be kept for a lifetime with a

---

* The text of this section derives in part from a feature article published under the same title in *Chemistry in Britain*, January 2002, pp. 30–32.

minimal metabolism and no significant turnover. Moreover, the transparency of the lens depends on the proteins' continued well-being. Mature eye lens cells contain a higher protein concentration than a solid protein crystal, yet the proteins stay in solution. If they started to lump together and come out of solution, as most proteins would do even at much lower concentrations, this would render the lens turbid (a condition called cataracts, which will be described in more detail below) and could lead to blindness.

What are these proteins like and how can they survive for so long? Of the extremely high protein content of the lens, around 90 per cent is made up of the typical eye lens proteins, called the crystallins. There is one group of crystallins that is universally conserved in every animal that has an eye with a lens. This group includes the α crystallins, which are believed to keep their equals and the other proteins from denaturing and lumping up to insoluble aggregates. Proteins doing this kind of job are called molecular chaperones, and have been a focus of protein biochemistry throughout the 1990s. The α (alpha) crystallins are indeed closely related to a group of heat shock proteins known to serve as molecular chaperones. The first group also includes the β (beta) and γ (gamma) crystallins, which biochemists at the University of Regensburg, Germany, and at Birkbeck College, London, have studied extensively with the aim of working out how evolution has engineered a protein of such remarkable long-term stability. Their structures were found to contain modules (domains) which apparently descend from a single module that was duplicated twice during its evolution. Both the structures of the individual domains and their mutual interactions appear to have been optimized to guarantee maximal stability.

The second major group of crystallins, following the Greek alphabet from δ (delta) through to σ (sigma), is equally unusual. Unlike the universal α and $\beta\gamma$ proteins, each of these proteins is specific to a group of organism. More intriguingly still, each of them has a day job down in the body, serving as some metabolic enzyme or another. When in the lens, these proteins are modified in a way that tunes down their enzyme function while enhancing their stability. But there is no doubt that, for instance, ε (epsilon) crystallin in the duck's eye lens is the product of the same gene as the enzyme lactate dehydrogenase found in the same animal's heart muscle. This unique multitasking has been dubbed 'gene sharing'. Presumably evolution has recruited whatever enzyme in a given species was stable enough to serve as an eye lens protein. But there is still a vague feeling that we may not fully understand what nature is up to.

But what happens when the crystallins fail to stay folded and solubilized? In that case, they form aggregates and the lens becomes opaque, a condition called cataracts. This can happen as a consequence of accumulated damage in old age, but there are also a number of genetic mutations that will lead to an innate tendency to suffer from cataracts in childhood or even at birth. In such cases, early diagnosis and treatment is of utmost importance, as the children would otherwise fail to build the brain structures required to make sense of the signals coming from the eye. One young patient with a genetically caused cataract was recently described as having accumulated protein crystals in his eye lens. While most cataract cases are believed to arise from aggregation of destabilized protein, it is perfectly plausible to assume that some of the mutations affecting crystallin solubility may result in crystallization of an otherwise stable protein.

The effects of cataracts on vision have been fairly well characterized since the nineteenth century. Prominent sufferers, including the painter Claude Monet (1840–1926), have described their loss of vision, which in the early stages can also manifest itself as a red shift in colour perception. Some art historians interpret the characteristic changes in the colours that Monet chose to paint his recurring motifs, including the famous bridge in his garden, in terms of the colour effects brought on by his deteriorating eyesight. In the initial stages, as the cataracts formed first in the oldest, central parts of the lens, Monet benefited from pupil-widening eye drops. He could effectively look past his cataracts as long as his pupils were wide enough. Eventually, in 1922, Monet had the opaque lens removed from his right eye. With strong spectacles, he could see normally with the operated eye, while the left eye still had a lens with some cataracts. This allowed him to mix and match his colour perception—more blue-green on the right eye, more red-yellow on the left.

An improved operation allowing cataract patients to see normally without strong spectacles only became possible several decades later thanks to a serendipitous observation, and the tenacity of a pioneering eye surgeon. During the Second World War, Harold Ridley (1906–2001) had treated fighter pilots whose eyes had been hit by splinters of the perspex material (poly(methylmethacrylate)) of which the cockpit windows were made. He noticed that the eyes tolerated the foreign substance surprisingly well—there was no sign of inflammation or particles moving around. At the time, dogmatists fiercely defended the view that an artificial lens would be impossibly difficult to implant, and would also be rejected by the immune system. Therefore, Ridley conducted his first implant of a perspex lens in 1949 in secret, and only published his method in 1951. He succeeded with this and subsequent treatments, but the method proved difficult for others and thus remained controversial for decades.

It was only after the introduction of improved instrumentation in the 1970s that the lens replacement became universally accepted. Nowadays, around five million people around the world receive this treatment every year. Although a few other materials including silicones and hydrogels have also been tried, perspex is still the most commonly used. However, the lenses don't come from the same manufacturing line as the average garden shed window. As any contamination even with particles as small as 10 micrometres might cause rejection of the implant, the acrylic sheets for surgical applications are normally made and processed in separate, purpose-built facilities.

Ridley himself benefited from 'his operation' as well, receiving lens implants in both eyes. Other benefits, including official recognition of his achievement, were extremely slow to materialize, however. He was only elected to the Royal Society in 1986, 15 years after his retirement, and narrowly escaped the fate of Charles Darwin—to die as a plain 'Mr'—when he was knighted in 2000, just a year before he died aged 94. (Surely the previous prime ministers had very good reasons to save their gongs for more important people, like that nice Mr Archer you may have heard of!)

Current efforts to improve the technology further are mainly aimed at making artificial lenses which not only sit there and fill the space, but which can change their shape and thus allow the patient to accommodate to different distances as they previously could with their natural lenses. Bifocal lenses with separate fields for near and far vision, working much the same way as bifocal spectacles, have been available since the early 1990s, but these still provide optimal vision only at two fixed distances. Researchers at Vanderbilt University (Nashville, Texas) are currently developing a new kind of composite lens that is designed to respond to the muscles which mediate the distance accommodation in the same way as the natural lens. The new-style implant will consist of six overlapping lenses which will overlap more (and refract more strongly) when squeezed by the muscle. Inbuilt wire springs will make the arrangement expand again as soon as the muscle relaxes. At the time of writing, clinical trials of the new implants were expected to begin soon.

Of course it would be even better to avoid the cataracts forming in the first place. This requires a detailed understanding of why they form. One influential school of thought is focusing on the effects of oxidation reactions brought about by active oxygen species generated by photochemical reactions involving blue and UV light. After all, the eye lens is exposed to light for as long as you keep your eyes open, and this adds up to a lot over a lifetime. One key observation is that ascorbate (vitamin C) is up to 30 per cent more concentrated in the aqueous humour than in the plasma (the cell-free part of the blood). This suggests that it may play a role in preventing oxidative damage to the eye.

Intervention studies with thousands of nutritionally deprived elderly people in China have suggested that vitamin supplements can indeed reduce the probability and slow down the development of cataracts. Apart from vitamin C, vitamins A (which also keeps the retina healthy) and E, thiamin, riboflavin, and essential elements are thought to play a part. As yet, physicians are far from agreeing on whether normally nourished people should take vitamin supplements to protect their eyesight, let alone how such supplements should be formulated. Many think that taking standard multivitamins within the recommended limits will be beneficial. In fact, there is a significant correlation indicating that people who say they take multivitamins regularly are less likely to develop cataracts.

The eye lens keeps intriguing everybody interested in protein stability. Every seeing person should be grateful for the amazing properties of the lens proteins, because if they behaved like normal proteins our eye lenses would have exactly the consistence and transparency of the white of a boiled egg.

Cells with a vision

At the back of your eyeballs is a layer of light-sensitive cells which—like the film in a camera—records an upside-down small-scale image of what you are seeing. The retina is much cleverer than a conventional film, as it not only catches the light and the information it carries, but it can also convert it into an electrical signal (essentially like a digital camera) which will then be transmitted to your brain and create a visual perception. Strictly speaking, the retina is a part of the brain, as the light-sensitive cells have evolved from brain cells. There are two kinds of light-sensitive cells in the retina. The cones are mostly found nearer the middle, they can distinguish colours, and they are less sensitive than the rods, which are colour blind and more abundant near the edges of the retina. A human retina contains roughly a billion rods and three million cones.

To find out just how a biological cell can convert a light signal to an electrical one, let us now consider the architecture of the rod cells (Figure 5.3). They are subdivided into two major segments. The outer segment (which ironically is turned inside, away from the light in the

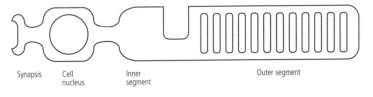

Synapsis   Cell   Inner   Outer segment
             nucleus   segment

Figure 5.3 Schematic cross-section of a rod cell from the human retina.

human retina for historical reasons) contains a pile of membrane 'discs' into which the molecules of the light receptor, rhodopsin, are incorporated. Roughly speaking, the outer segment does the signal conversion job, while the inner segment keeps the cell alive and contains all the compartments and structures that are needed for the life of any eukaryotic cells, including the nucleus, the mitochondria, and lots of ribosomes, as the rods are very active in protein synthesis. At the far end of the inner segment there is the synapsis which provides the link to the nerves connecting to the brain.

One of the amazing things about rod cells is that each one can be stimulated to produce a nerve impulse by the arrival of a single photon (the smallest possible amount of light according to quantum mechanics). The primary chemical reaction triggered when a photon hits the cell happens in a molecule called retinal, which is a slightly modified version of vitamin A and has to be bound to the membrane protein opsin, which is called rhodopsin for as long as it carries retinal. Within a picosecond (10^{-12} seconds) after the arrival of the photon, the kinked structure of the chain-like part of retinal changes to a stretched-out version. In a complex series of consecutive reactions, the rhodopsin will now activate a specific signalling protein (transducin), from which the signal will be passed on and amplified by several orders of magnitude, until it eventually reaches certain ion channels which will briefly block the way for sodium ions. This results in a rapid change in the electrical potential across the membrane of the outer segment—in other words, a nerve impulse is born.

All this happens within a second when just one photon arrives, and even faster for a more intense light pulse. For the record, I should note that the straightened-out retinal does not fit its binding site in the opsin protein any more and will be released from it in a slow reaction. It will have to undergo a chemical reaction which restores the kinked form before it can bind to an empty opsin molecule again.

However, the function of the rod cell is not only something that happens on the protein and chromophore level. Membranes play a very important part, too. The rhodopsin is actually incorporated into the peculiar membrane discs stacked up in the outer segment of the rod cell. The way it gets there is curious, in that the membrane protein first gets incorporated into the plasma membrane of the cell, which then bulges inwards to form the membrane bag that will flatten out to a disc. It can be shown that the sugar molecules attached to rhodopsin are at first on the outside of the cell and then end up inside the disc structure. From the distribution of the hydrophobic (water-avoiding) amino acid residues in the opsin sequence, researchers have derived hypothetical models of how the protein chain winds its way back and forth through the membrane in seven transmembrane helices. However, rhodopsin is among the most important proteins for which no high-resolution structure has been solved so far.

One intriguing aspect of the function and metabolism of the rod cell is that the energy required for restoring the active, kinked version of the retinal chromophore is not provided by a small molecule such as ATP (as one would expect) or GTP. Instead, the very molecules which make up the membrane into which the bacteriorhodopsin is embedded also provide the energy. As Robert R. Rando's group at Harvard has shown, the energetically more favourable stretched version reacts with a membrane component called lecithin to form an energy-rich compound known as an ester. The enzyme that restores the kinked structure uses the chemical energy stored in the ester bond (which it cleaves) to introduce the kink.

Considering the wealth of differently constructed eyes, it is note-worthy that the rhodopsin/opsin mechanism has been found in every example that has been investigated. It is also used in our second set of photoreceptors, the less numerous and sensitive, but colour-specific cones.

Colour vision puzzled Victorian scientists profoundly, and even John Dalton (see FLASHBACK on p. 120), who was the first to describe the relatively frequent hereditary colour blindness in objective terms, failed to understand its mechanism. Even today, scientists are still struggling to combine the findings of physiological and psychological experiments into one definitive description of how colour vision works.

The primary detection is quite simple. Most people have three

John Dalton (1766–1844) and colour blindness

C HEMISTS KNOW JOHN DALTON as the creator of the modern concept of an atom, and for his work on how partial pressures of gases combine. Ophthalmologists, however, associate his name with the phenomenon sometimes called Daltonism, but more widely known as colour blindness.

Dalton, the son of a weaver, owed his profound education to his Quaker background. He entered science with the publication of his meteorological observations in 1793. In the same year he moved to Manchester, where he stayed for the rest of his life and made a living from teaching and consulting. The physicist James Prescott Joule was among his pupils. In 1808 and 1810 he published the first and second instalments of his *New system of chemical philosophy* which, by defining atoms as the unit of accounting for chemical reactions and attributing them relative weights, laid the foundation of modern chemistry.

Here, however, we are mainly concerned with his investigation of colour blindness. That such a phenomenon should need to be 'discovered' as late as 1794 is an intriguing idea at first sight. On second thoughts, however, one realizes that before the arrival of industrial dyes and electrical lights, colours were far less present and important in daily life. Furthermore, analysing a phenomenon which is only present in certain people's perception requires a mind working with strict scientific method such as Dalton's. Although a few observations had been made of people who appeared to have difficulties distinguishing colours, no analysis of any scientific rigour was undertaken before Dalton.

Although he had sometimes noticed that he couldn't agree with other people on the naming and similarity of colours, Dalton only got involved more deeply in this issue when, from 1790 onwards, he developed an interest in botany. Trying to identify plants by the colour of their flowers or fruit, he noticed that what others called blue, purple, pink, or crimson would all be indistinguishable from blue in his perception.

In 1792 he was struck by the observation that a (pink) flower which he had seen as 'blue' in daylight, now appeared 'red' to him when he saw it in

candlelight. He asked several people whether they
could observe a similar change in colour, but none could,
except his brother. Systematically checking through his perception of the colours of the
spectrum against other people's, he found that he could only distinguish two or three
colours where others saw six or seven.

From one of the few previous mentions of a person unable to distinguish colours,
he tracked down another family with the same anomaly, where four out of six brothers
were colour blind. Testing school classes he found an incidence of one to two pupils per
class. Today the incidence is about 8 per cent for males and 0.5 per cent for females.

John Dalton laid down a solid foundation for future studies of the phenomenology
and distribution of his condition—all without the slightest chance of imagining what
vision might be like for the majority of people who can see the full colour spectrum.
His willingness to consult and trust other people's perceptions was part of what made
his observations quite unusual for his time. Today's ophthalmologists are still
impressed by these studies and have no problem pinning down a precise diagnosis
from what Dalton reported about his 'case' in a paper he read to the Manchester
Literary and Philosophical Society on 31 October 1794 ('Extraordinary facts relating to
the Vision of Colours with Observations').

What he got quite wrong, however, was the interpretation of the phenomenon.
He supposed that the liquid in his eye might be coloured blue. At the time, this
explanation wasn't wildly implausible, but in 1801, when the British physicist and
physician Thomas Young (1773–1829) pointed out that all colours of the visible light
can be produced as a mixture of just three components, suggesting that only three
different kinds of colour receptors are required in the eye, he could have abandoned
this view. Young suggested that the absence or malfunction of those parts of the retina
designed to see red light caused the specific colour blindness experienced by Dalton.

In 1830, after the astronomer John Herschel (1792–1871) had carried out colour
vision experiments with Dalton as a subject and come to the conclusion that his vision
did not lack in intensity at all, but was explicable with reference to two receptors
instead of three, Dalton still persisted in his belief. Posthumously, his eyes were
scrutinized and found to be normally transparent—and not at all blue.

different kinds of cones, equipped with visual pigments specific to a certain wavelength range. Specialists call them S, M, and L cones (for short, medium, and long wavelength), but there is nothing wrong with calling them blue, green, and red cones. At a molecular level, they all work with retinal (and the kinked versus straight mechanism described above), but the wavelength range of the light absorption is shifted by the influence of the protein component.

The distinction between green and red cones is a relatively recent invention in our evolution, and we only share it with our nearest primate cousins. It is thought that the ability to spot edible young leaves and/or fruit provided the evolutionary benefit which helped us to become trichromatic. Around 2 per cent of the male population, however, lack either red or green cones and are therefore only dichromatic (the term colour blind is a better-known but somewhat inadequate description of this condition—the total inability to see any colours is extremely rare).

Note that there is nothing magical about having 3 cone types. Other mammals get by with 2, while most birds have 4, and some crustaceans up to 12. Amazingly, decades of research into the use of colour in sexual selection in birds failed to take into account that they see the world of colours differently from us. It was only when researchers began to look at the UV reflectance of their plumage that they realized that male and female blue tits, though nearly indistinguishable to the human eye, look very different to a bird's eye.

Things get more complicated and controversial when the signals transmitted by the cones start out on their voyage to the vision centre of the brain. The current view is that the three kinds of signals feed three different kinds of neurons, which compute them into a quite different set of variables comprising a red-green, a blue-yellow, and a brightness channel. The red-green channel, for instance, calculates a difference between the signal from red and from green cones (L-M).

Whether these are the only channels, and in which ways exactly their signals lead to the perception of colour that can be monitored in psychological experiments, remains to be uncovered because, as I will explain in the next section, what we perceive is sometimes fundamentally different from what our eyes see.

Why your head is *not* a camera

The eye has optical components reminiscent of a camera: the light focused by the cornea and the lens will create an upside-down image on the retina, much like the image captured on film in a camera. So it would be really tempting to drive this analogy further and say the retina captures so many pictures per second and sends them to the brain so we can see the movie of what's going on before our eyes. However, life turns out to be a lot more complicated, as we shall see in due course. We shall try to work out the difference between what we see and what we perceive, and also discover strange things going on at the borderlines of conscious and unconscious vision processes. And by the end of this chapter you will probably be glad that your head is not a camera.

Simple optical illusions, often requiring no more than black lines on white paper, are all you need to demonstrate that what you perceive (the movie in your mind) is different from what your eyes see (the movies on your retinas). In Figure 5.4, the vertical lines in (a) are of equal length, and so are the two horizontal lines in (b). Their images on the retina will be of equal length. And yet most people will perceive them to be different.

Some of the classical optical illusions have been studied by psychologists since the early twentieth century and explained in various, often contradictory ways. A consensus that has emerged more recently is that most of the illusions are due to three-dimensional interpretations of the two-dimensional pattern seen. Thus, it is said that the deception in Figure 5.4(a) and (b), based on the three-dimensional interpretation of the straight lines, works less well in people from the Zulu culture, who only know round buildings and very few linear structures.

A robot equipped with a video camera would not be fooled by images such as Figure 5.4. But then the same robot might have problems identifying objects when they are partly hidden, or seen from an unfamiliar angle. The unconscious interpretation work that our brain applies to visual data coming in is a major advantage in such situations. You know, for instance, that tables and chairs tend to turn up together. Hence, if you see the back of a chair behind

(a)

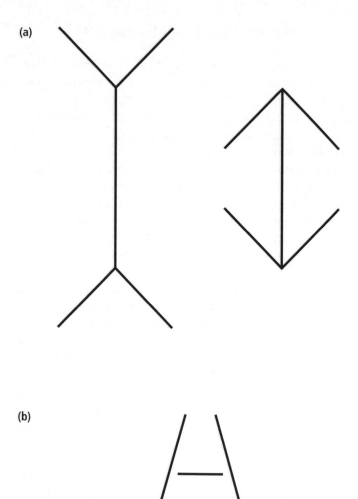

(b)

Figure 5.4 Two simple examples of optical illusions. In contrast to their appearance, the vertical lines in (a) are of equal length, as are the horizontal lines in (b).

a table, you immediately interpret this visual impression as a complete chair being partially hidden from your view by the table. (You wouldn't consider the possibility that somebody could have mischievously nailed part of a chair to the side of the table.)

So the mental processing which fools us when we look at illusions is important for recognizing objects. Another reason why we interpret and extrapolate what we see is that we cannot view all directions at the same time. Although at any given moment we can see only a small part of our visual field in clear detail, we do have a visual impression of the whole surroundings. This impression is generated by rapid eye movements, called saccades, which we perform unawares up to three times a second.

Researchers originally thought that a concrete image of our surroundings is stored in the brain, which in a particular moment is updated for the particular direction of our gaze. Experiments in which the visual field of a person was changed rapidly during the saccadic movement, however, proved this assumption wrong. It now appears that the data storage device which allows for continuity between images before and after a saccade and creates the illusion of a continuous 'movie' representing our surroundings, is much less detailed and concrete than originally thought.

This so-called trans-saccadic memory seems to record important landmarks in a rather schematic way in order to allow us to make the link between successive views. Details are not recorded and are left for the conscious gaze in a particular direction. Psychologists have, for instance, asked people to read texts consisting of aLtErNaTiNg upper- and lower-case letters on a computer screen, while their eye movements were recorded. During a saccadic movement, the researchers could switch all lower case to upper case and vice versa without affecting the subject's reading speed and even without the subject noticing any change. Obviously, the trans-saccadic memory regarded the upper/lower case as insignificant and did not record it at all.

Such 'change blindness' has also been observed with people looking at pictures or, indeed, watching movies. As the scenes of a movie are not normally filmed in the same order in which they will eventually appear on the screen, there is always one person on the set (the continuity assistant) who is in charge of consistency of optical details. Yet it can still happen that the hero of a movie boards a car in jeans and gets out in a suit. Such violations of logic and continuity are contained in most films, but most of the time nobody even notices them. Experienced film-makers know how to cut in a way that navigates viewers around the inconsistency. If they keep looking at the actor's face, for instance, they will not notice the changing trousers.

Change blindness can even affect the object on which a person's attention is focused, as a surprising experiment with computer graphics has shown. Subjects were asked to move coloured squares around on the screen using a mouse, to reproduce a given pattern. Naturally the eyes move between the representation of the task, the objects to be moved, and the target area. If psychologists changed the colour of one of the squares during such an eye movement, the change went unnoticed in most cases.

Such experiments, along with the experience from the film industry, suggest that the brain does not keep a detailed video recording of each moment—not even for the duration from one glance to the next. But what about our long-term memory of scenes and pictures that we have looked at for several seconds? We tend to think that we can remember certain visual scenes in every detail. People shown hundreds of photos can often tell with more than 95 per cent accuracy whether they have seen a given photo before or not. This looks impressive at first glance, but then, if psychologists get up to their mischief and start changing details of the pictures or showing mirror images of them, that does not affect the classification as known or unknown. Like the trans-saccadic memory, the long-term memory also keeps a schematic view of a given scene, rather than a complete copy containing every pixel. The photographic (eidetic) memory that most children have is normally lost when they grow up, because in life it's not the pixels that count, it's their meaning, and that can be stored in a more efficient way than by photographic memory.

So the movie on the retina is only the crude material which our brain uses to create the movie in the mind. But how is this processing done; where is the crew? Using modern brain-imaging techniques (mainly those based on magnetic resonance imaging, or MRI), scientists can work out which areas of the brain are involved. But the question of what actually happens is muddled up by the fact that both conscious and unconscious processing steps are involved in the whole perception pathway. Luckily, recent research is beginning to unravel the two.

How to see without seeing

Now we have seen that what we consciously perceive is different from the image on the retina. But is there a part of us that sees the retina movie without us being conscious of it? And if so, how can conscious and unconscious vision be separated?

One important clue in this quest comes from patients who have been unfortunate enough to lose sight in part or even the whole of their visual field through injuries or strokes affecting the brain regions at the back of their heads which are involved in the processing of vision, called the visual cortex. Such people have fully functional eyes (the images on the retina are still there) but they are partially or completely blind due to the breakdown of the neuronal processing of the images which lead to conscious perception in a healthy person. As this processing is blocked by damage to the cortex, the condition is also called cortical blindness.

Ernst Pöppel and his group at MIT first reported that cortically blind people can respond unconsciously to visual stimuli. Larry Weiskrantz and his coworkers at Oxford made similar observations and coined the term blindsight in 1973. Although such individuals are not aware of seeing an object, for instance, they will, if forced to make a guess, identify it correctly more often than could be explained by chance. Or they can shape their fingers to the right kind of diameter to pick up a ball that they cannot see.

Some media reports have described blindsight as a freak phenomenon found in very few people with specific brain damage. But Petra Stoerig, who is also investigating this phenomenon with her group at the University of Düsseldorf, Germany, says that cortically blind patients, who initially may not show this ability, can be trained to develop and use it. For people who have lost half of their visual field or more, such training, which is as yet an experimental procedure and not part of the established therapy for stroke patients, can make all the difference between bumping into the furniture all the time and being able to navigate around the room like a seeing person. Stoerig believes that anybody suffering vision loss through cortical injury could be taught to benefit from blindsight. She is currently conducting clinical trials to get this approach accepted as a therapy. Another important task ahead of her is to spread the awareness that such problems originate in the brain rather than in the eyes—patients who lose part of their visual field in a stroke often don't even know that the specialist they need to consult is not the optician but the neurologist.

Another indication that blindsight is a general phenomenon comes from an elaborate perception experiment carried out by Christopher Kolb and Jochen Braun at Caltech using healthy subjects. In a screen display showing lots of small lights moving back and forth over short distances, the subjects were asked to identify a target area where the lights moved differently from the bulk of the screen, and then to rate their confidence in the reply. In the first experiment, where the dots were clearly visible, confidence ratings were in good agreement with the success rate in identifying the target area. In a second experiment, however, the psychologists obscured the movement of the dots by pairing each of them with a second dot moving in the opposite direction. The conscious perception sees nothing but flashing lights on the screen. If forced to guess, however, the subjects still identified the target area with roughly the same success rate, but with a uniformly low confidence rating that no longer correlated with the actual success.

Thus it appears that, apart from the multistep visual processing pathway that leads to the conscious perception—the movie in the mind—there is a different, subconscious pathway that can also guide our actions and guesses, but never reports its observations to the conscious mind.

Another way to dissociate vision from conscious perception is easier to achieve and has puzzled scientists for more than two centuries. It relies on the fact that our eyes normally see somewhat different but overlapping views of the world, which can be merged into one image by the neuronal processing in the cortex. Now if by some trick (a stereoscope or an empty kitchen roll can be used for this) the two eyes are presented with two separate, incompatible images, what happens? Will the perceived image be a mixture of the two? Or will they be presented side by side?

As you can easily find out for yourself, none of these things happen. At least for most observers, a brief period of confusion is followed by an alternation of the two competing images at apparently random time intervals. The French physicist M. Du Tour described in 1760 how if he placed a piece of cardboard with a yellow patch on one side and a blue one on the other alongside his nose, he never saw a combination image of a green patch as he might have expected, but always either the yellow or the blue one. This phenomenon is known as binocular rivalry and has since been investigated and interpreted by many other scientists, including the German physicist Hermann von Helmholtz (FLASHBACK, p. 130).

Binocular rivalry (see Figure 5.5) is an ideal system for distinguishing between conscious and unconscious vision, because you have both happening in the same individual at the same time. Each eye has its own movie, and the visual brain gets transmissions of both, but the conscious perception keeps (unconsciously) zapping between the channels, watching each for a few seconds at a time. Nikos Logothetis and his group at the Max Planck Institute for Biological Cybernetics at Tübingen, Germany, have used this approach to investigate the electrical response in individual nerve cells in monkeys trained to report which of the conflicting images they were perceiving.

The picture they obtained was somewhat muddled at the beginning of the processing pathway, but thankfully became clearer further on. In some of the brain regions that get a visual signal first, as few as 10 per cent of the neurons were activated if the stimulus supposedly more efficient for this brain region was on the active channel. In some areas, a comparable percentage of neurons responded to the stimulus only when it was not perceived. Whether this is a part of a pathway towards a neuronal dustbin ('I don't want to see this!') is not known so far. The most specific response was

Hermann von Helmholtz (1821–94) and vision

A T A JUBILEE CELEBRATION IN BERLIN IN 1891, one of the most famous nineteenth-century physicists, Hermann von Helmholtz, expressed his modesty and embarrassment at all the honours heaped upon him on the occasion of his seventieth birthday. Trying to keep his feet on the ground, and not let the fuss made about him go to his head, he recalled his modest beginnings. His 'poor memory for unrelated facts' drove him into trying to find relationships such as physical laws which would make things appear simpler and easier to understand and remember. And his interest in physics, which wasn't really considered a profession that one could make a living by, was channelled into the studies of medicine for purely economic reasons.

However, his search for rational relationships, combined with his background in both physics and medicine, made him one of the most prolific scientists of the nineteenth century. In physical chemistry he contributed fundamental concepts such as the free energy, and also made important progress in fluid dynamics and electrodynamics. He is also credited with laying the foundations of musical acoustics as a physical science.

Helmholtz was ideally placed to begin applying physical methods to medicine. Thus he was the first person to measure the speed of propagation of a nerve impulse. Independently of Charles Babbage, he invented the ophthalmoscope (see p. 109) and thus became the first person to see a living retina. Vision was one of his main interests among the biological applications of his research. At the time, large parts of this field were still open to speculation and debate, and thus a famous controversy between Helmholtz and his coworkers on one side, and the school of Ewald Hering on the other, regarding the understanding of vision and perception, was carried on for decades, well into the twentieth century.

Two key issues on which the two schools disagreed were perception of three-dimensional space and perspective, and colour perception. Hering emphasized the observation of three channels (red-green, blue-yellow, and luminance), while Helmholtz's analysis focused on the detection of the three primary colours: red, green, and blue. Today, both views are reconciled in that the primary-colour receptors have been identified as colour-specific

cones, while the three channels are regarded as the result of the way in which the neurons compute the signals they get from the cones.

Helmholtz was also among the first authors to describe and analyse the phenomenon of binocular rivalry, described on p. 129. He studied the phenomenon using rivalry between pictures presented to one eye and printed words to the other, in which case it is relatively easy for the subject to choose which of the perceptions should prevail, by either trying to read the words, or trying to trace features in the picture. Helmholtz noted that the perceptual switch cannot operate by a decision for one eye. Rather, the individual needs to have a clear expectation of what they are going to see with that eye, suggesting a relationship with the mechanism involved in guiding visual attention.

From his observations of binocular rivalry, Helmholtz concluded that the images received by both eyes don't simply add up to a three-dimensional impression, but that their recombination must be an act of mental processing. It's only now, more than a century later, that we are getting the first glimpses of how this mental act might work.

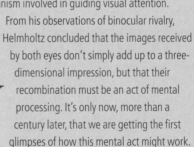

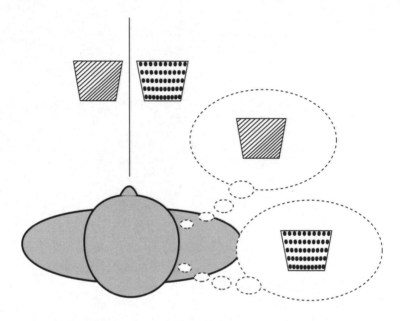

Figure 5.5 Binocular rivalry. When the eyes see incompatible images, perception switches between the two at irregular intervals rather than mixing them.

observed in the inferior temporal cortex (ITC), where around 90 per cent of the neurons responded specifically when the preferred stimulus was perceived.

What is not quite clear, however, is the working of the (unconscious) remote control that does the channel-hopping. To some extent, one can influence the switching consciously. For instance, if one eye is presented with a text and the other with a pattern, one can bias the rivalry by trying to read the text, or by trying to trace the pattern. If such control is not attempted or not possible, however, the time sequence of switching is indistinguishable from a random pattern.

Combining such approaches to the conscious and the unconscious processing of visual information reaching our brain with modern and still improving imaging techniques such as MRI will continue to improve our understanding of the visual process, and how the conscious and unconscious parts of the brain interact. One big challenge ahead is to understand what's happening in the situation exactly opposite to blindsight, when you perceive a movie in the mind without having a movie on the retina. This strange phenomenon is very poorly understood by science, but quite commonly known as dreaming.

How to be seen, and how not to

In biology, there are at least two parties to everything: one creature eating means another one being eaten. And the story of vision would not be complete without a mention of what happens at the other end of the light beam, that is, what measures organisms take in order to be seen or not to be seen. Some of the obvious measures have been discussed in Chapter 3: bioluminescence can be used either to be seen by potential mates (as in insects), or to appear less visible to predators, as in the fish that mimics the daylight to disrupt its own silhouette, or the squid that sprays clouds of glowing particles to cover up its flight.

One efficient way of remaining invisible in the ocean, where hiding places are scarce, is to be transparent. This strategy is widely used by squid, jellyfish, and crustaceans, and it is not as strange as it sounds. If you live in the dim light of the underwater world and don't practise photosynthesis, you have very little reason to carry strongly light-absorbent substances with you. And it's a small step for evolution from a colourless animal to one that is transparent to the degree that it is hard to spot in the water. There are only two fundamental problems with this approach. Firstly, the content of a transparent animal's stomach may be visible. And, secondly, if the creature in question has eyes, these need pigments that absorb light, and thus would appear dark as a consequence of their very function. Animals that depend on their transparency for camouflage reasons often have very small eyes or very large ones. If all the pigments are concentrated in one spot, they might be overlooked or considered a speck of dirt. If they are spread out over a large area, they may add so little to the absorbance that they could go unnoticed as well. Some predators, however, can still see creatures that appear transparent to our eyes, and it is thought that these perceive changes in the polarization of the light, which we cannot detect.

The other possibility of making oneself visible or invisible is camouflage by using colour and pattern. While some innate similarity to the background that an animal is most likely to live on may already be an advantage, the most efficient camouflage is obviously the one which can directly and quickly match a creature's skin to the current background. Various fish, frog, and reptile species can do this, often to stunning precision. In a recent study into the rapid background adaptation of a tropical flounder species, for instance, researchers went to great lengths to ascertain that the stunning pattern-matching abilities of the fish were real, and not due to the organism simply being transparent.

But how do these animals manage to change their appearance within seconds? It is an engineering challenge similar to the task of creating a newspaper that rewrites itself every morning. You want to change the darkness of the surface of a rather thin sheet (the paper or the skin) with good local precision. Essentially, as unpigmented skin (like paper) would be white, you want to have pigments and you want to move them around. For the newspaper, someone came up with paper containing a half-black, half-white ball for each pixel. Opposite electrical charges located on the dark and the light side allow the ball to be turned around by electrical signals.

Nature's method is a little bit different, but also relies on little pigment balls. They are called melanosomes and are black all around. Hundreds of them are in each of the colour-changing skin cells (melanophores). Such a cell may be told to get lighter by a hormone signal—one of the possible messengers is the hormone melatonin, which we met in the last chapter in its role as a regulator of the circadian clock, but which owes its discovery and its name to the effect it has on frog melanophores. Upon receiving this signal, the skin cell accumulates all its melanosomes near the centre, like the morning rush hour sucking thousands of commuters into a big city. Here, the melanosomes are more likely to cover each other, so the overall impression of blackness will be reduced. Conversely, if the cell wants to appear blacker, it actively transports the commuting pigment balls back to the suburbs again.

There is one intriguing detail that I can't help mentioning, although it will probably interest only biochemists: transport into the city is carried out by a different railway company from the one that shuffles the melanosomes back out to the suburbs. On the way in they travel on microtubule rails using dynein trains. On the way out it's either microtubules and kinesin-II trains, or actin rails (as in muscle contraction) using myosin-V trains. This kind of separation makes more sense in the cell than it would in the commuter's world, as it allows the cell to switch on one direction with high selectivity by powering one kind of motor or the other.

Finally, there may be third parties to this hide and seek game. This is when evolutionary games get really complicated. Butterflies may, for instance, avoid being eaten by mimicking the patterns of poisonous species. In the extreme, this can provide a significant advantage for the impostor, at the cost of the unpalatable species being mimicked. The presence of edible mimics waters down the predator's learning effect and thus puts the inedible 'original' at a higher predation risk than it would have in the absence of the mimic. It emerges from recent research that such relationships can be very complex indeed, as the behavioural adaptations of several species have to be taken into account.

The limits of light

The performance of light as an information carrier is quite impressive, both for its speed and for its efficiency. There are, however, limits to what it can do, which are dictated by the fundamental laws of physics. These are reached when the size of the object in question is comparable to the wavelength of the light, that is 400–700 nanometres (millionths of a millimetre). Ironically, this is exactly the length scale on which the main components of our cells operate—the proteins, nucleic acids, and complex assemblies of these (Figure 5.6).

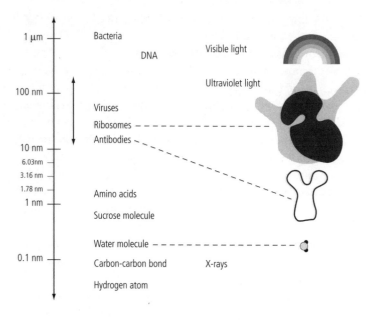

Figure 5.6 Natural objects on length scales smaller than the wavelength of visible light. Note that the scale on the left is logarithmic, i.e. every major tick indicates progression by a factor of 10.

Thus, whatever we do, we shall never be able to see a protein as we can see a cell under the microscope. If we are lucky, we can create images of such nanometre-scale structures using radiation of much shorter wavelength. This is the underlying principle of both the electron microscope, which can take pictures of individual molecules down to a scale of around 1 nanometre, and of X-ray crystallo-graphy, which operates at even smaller lengths, but which requires thousands of identical copies of the structure in question to be arranged in a regular crystal. The trouble with X-rays is that so far there are no lenses or mirrors for them with which to build an X-ray equivalent of a microscope that could look at individual units rather than crystal patterns. Scientists at the Rutherford Appleton Laboratory near Didcot in Oxfordshire are using one of the most powerful lasers in the world to power an X-ray laser which may one day allow us to see nanoscale structures with X-rays. And, as I am writing this chapter, the news is coming in that good old vinyl LPs can be used as X-ray mirrors. So don't throw them away and watch this space …

Both X-ray crystallography and electron microscopy essentially give three-dimensional maps of electron density. How we translate these into something that our eyes, trained on macroscopic objects, would recognize as a representation of shape is purely a question of convention. You can 'see' atoms as dots or balls on sticks or balls merged into each other; it doesn't change the fact that we cannot really see atoms, but only imagine them.

I have discussed the workings of the nanometre-scale world in some detail in a previous book (*Travels to the nanoworld,* 1999), so I don't want to repeat myself too much here. One instructive example from that book, however, may merit at least a pointer. Semiconductor particles of nanometre dimensions have weird properties governed by quantum mechanics and are therefore known as 'Q particles'. (Among other things, they can be used to create what's known as quantum dots.) One of the surprising effects observed with Q particles is that their apparent colour depends on their size. In the macroscopic world one can be pretty sure that cutting a red lump of material in two halves won't make it go blue. In the quantum world, however, this can happen, and it is again to do with the fact that the particle size is on the same order of magnitude as the light we use for vision.

This again highlights how subjective the term 'vision' is. What is visible and what isn't is defined by the light available, by the diffraction and absorption properties of the object in question, and by the properties of the organ that sees, and therefore by the physical parameters of the eye. All eyes that we know operate more or less in a wavelength range from near UV to near infrared, staying in the neighbourhood of our own window on the world. Even within that window, different sets of colour receptors can show you very different pictures, as the comparison between bird and human vision has shown. But for all we know, people inhabiting the fourth planet of Alpha Centauri may have eyes that detect X-rays. Thus, much as life depends on light, our definition of light, conversely, depends on life.

CHAPTER 6

Changing ideas about light and life

WHEN WE SAY 'THE SUN IS RISING', we echo the pre-Copernican worldview of a stationary Earth with the Sun moving around it. Similarly, expressions like 'to look at something' or 'feel somebody's stare' reflect the ancient Greek theory that vision is based on rays emanating from the eye. These are reminders of the fact that the theme of light and life is much older than science and has been thought about for millennia. Sun myths are consistently present in the religious and cosmological beliefs of most, if not all pre-scientific cultures, and the mysteries of vision and perception have intrigued philosophers of all eras. Thus, many of the phenomena described in this book have been described and explained again and again throughout history, and often very differently from the way we interpret them now.

Except for the flashbacks (which are meant to lighten the fact load rather than provide a scholarly treatise of the relevant history of science), I have thus far drawn a stationary picture of how late twentieth-/early twenty-first-century science describes the interactions between light and life. While this description is true to the best of our knowledge, different explanations may arise in the future and may have been given in the past. I would now like to emphasize that each scientific description of our natural world is a child of its time and may again be replaced by more comprehensive new paradigms.

So let's travel in time and space to collect some specimens of earlier thoughts about light and life. Starting in ancient Egypt (one of the most light-addicted cultures our planet has ever seen), we shall tour the globe in search of representative cultural manifestations of Sun worship, have a look at the history and science of eclipses, and then end this chapter with a glance on early ideas about colour, vision, and perception.

Sun gods through the ages

We may go 'worshipping' the Sun on our summer holidays, but once we are back in our everyday lives most of us would describe the importance of our central star in more rational terms. Science tells us that the Sun has three main functions: it is the centre of gravity of the Solar System (small wonder, as it contains 99.9 per cent of its mass), our most important energy source, and the light source that enables photosynthesis, not to mention vision. These facts are so well established to us that they verge on the trivial, but none of these three would have made any sense for people living before the time of Newton. And yet most ancient cultures appreciated the role of the Sun as provider of light and warmth, and often expressed this in religious terms, worshipping Sun gods who were attributed life-giving faculties. In some cultures the special attention given to the Sun and the seasonal changes of its path can be rationalized in terms of the crucial timing of agricultural chores. In others, it is just a symbol for light, power, and supernatural forces.

The role of the Sun in ancient Egyptian religion gives an impression of how the complex connections between light and life were recognized five millennia ago and interpreted in mythical or religious terms. While the emphasis shifted over the centuries, the Sun remained an important issue throughout. The lives and ideas of ancient Egyptians comprise an extremely wide field about which many books have been and many more could still be written, but I will restrict myself here to the briefest possible reference to their Sun worship inasmuch as it is illuminating for the present topic.

Egypt

In the earliest surviving glyphs from ancient Egypt, the Sun and Moon were seen as the two eyes of the main deity Horus, who was depicted as a falcon. In later, more complex belief systems, the Sun became the eye of the Sun god Re (or Ra). Humanity was believed to have sprung from the tears of this eye, while light was its glance. This interpretation is reminiscent of the ancient Greek belief that the vision process involves sending out light from the eye.

The relative importance of Sun worship in the religion, world-view, and politics of the ancient Egyptian empires may have changed somewhat during the millennia of its history. Nevertheless, ancient Egypt was the most stable civilization that our planet has ever seen, and worship of the Sun was always a part of it. Beginning with the second dynasty (2800–2675 BC), the name Re showed up in the names of pharaohs, and from the fourth dynasty onwards (c.2670 BC) pharaohs called themselves the sons of Re. The steps of which the first pyramids were made were meant to allow the deceased pharaoh to ascend to the Sun.

During the chaotic late phase of the Old Kingdom (fifth dynasty, from 2475 BC), Sun priests came to power, declared Re the main god, and had huge temples built for him. This, however, failed to stop the decline of the empire, which gave way to civil wars and local power centres.

From that time through to the New Kingdom (1550–1080 BC), Heliopolis, the Sun city (north-east of Cairo), was an important cult centre. Obelisks were regarded as symbols of Sun worship. Apart from Re, other Sun-related deities from the early pantheon of Egyptian religion still played important roles. Atum, the deity of the setting Sun, was credited with the creation. Atum is also representative of the Sun that travels through the underworld at night to re-emerge in the east the next morning. The concept of repeated renewal and rebirth apparent in the sunrise was central to the Egyptian religion. From this the pharaohs derived their claim to immortality. Like their ancestor the Sun, they were bound to rise again.

Atum was also in the centre of Egypt's only brief flirtation with monotheism. It was started by the 'heretic' pharaoh, Akhenaten, who ruled from 1364 to 1347 BC and was also famous for being married to that beauty queen, Nefertiti. He tried to remove all other deities and have only Atum worshipped. More than any previous Sun worshipping, the Atum of Akhenaten's worldview is identified with the notion of light more than with the solar disc.

At Amarna, in middle Egypt, he started building a new religious centre, after which his creed is known as the Amarna religion. While the traditional solar religion described the phenomena of the daily and yearly solar cycles with the help of iconic symbolism, the Amarna religion emphasized the importance of light for all living creatures. It produced hymns that described the response of living beings to the light and interpreted it as an act of worship. A hymn to the morning Sun includes the lines:

> Plants and trees move at the sight of you,
> Fish in the water leap at your appearance.

In spelling out the influence of the Sun on vegetation, the practical side of Sun worship is emphasized over the cosmological, world-order aspect that was dominant throughout most of the three millennia of Egyptian history. The old worldview, however, came back to power in a major backlash as soon as Akhenaten was dead and mummified. In Christian Jacq's widely popular Ramses novels, you can still see the reverberations of this change in the shape of a few clandestine Atum worshippers who want to restore power to Akhenaten's descendants, but who only constitute one among the many political conspiracies of Ramses' time (1290–1224 BC).

It is tempting to speculate that practical reasons for keeping an eye on the Sun may have contributed to the beginnings of the Sun religion and preceded the cosmological stories grafted on top of it. Ancient Egypt is an obvious case where the observation of annual cycles was of crucial importance for the success of food production, as the entire agriculture depended on the flooding of the Nile. In northern Europe, where the Celts built major structures including Stonehenge, the relatively short summers may have required that the agricultural calendar be kept with a precision that could only be derived from observation of the Sun.

Incas

Sun worship was also performed four millennia after the end of the Old Kingdom, on the other side of the globe, in the Inca empire in north-western parts of South America, roughly corresponding to what today is Peru and the northern parts of Chile. Unlike the Egyptian civilization, which matured and decayed over millennia, the Inca empire was less than two centuries old when the conquistadores wiped it out at a point when it was weakened by internal conflict. Only the kings of the New Empire, who called themselves Inca, picked the Sun god Inti from the multitude of traditional deities and declared him to be the patron god of their reign. While all South American cultures had some kind of Sun worship, the immediate predecessors of the Inca in the lineage of Andean cultures, the Chimœ, ranked the Moon deity higher than the Sun, and celebrated solar eclipses as a victory of the Moon.

And unlike the Egyptians, the Celts, and the Maya (whose main crop, maize, needs a lot of sunshine), the Andean people had no compelling reason to calculate or predict the Sun's path. For them the Sun represented immortality and the highest power. Their kings spread the belief that they were direct descendants of Inti and thus immortal like him.

According to this ideology, deceased regents lived on as mummies, and their property could not be passed on. Therefore, their successors on the throne had to conquer new land to be able to pay for their expenses at the royal court. While this kind of Sun worship worked well in the short term, it had direct and in the long term very damaging consequences on the political development of the Inca empire. The Incas were forced to expand fast and aggressively and thereby, their impressive infrastructure notwithstanding, they became vulnerable and quickly succumbed to the small group of conquistadores.

Eclipses

When considering the ancient thinking about the Sun, it is very difficult to draw a line between mythology and fact. But one kind of astronomical event that has inspired people to look at the celestial bodies in a scientific way and which has been recorded for at least five millennia are the eclipses of the Sun and the Moon. And there are very good scientific reasons for doing so, because until very recently the cyclic repetition of eclipses provided the most precise ways of measuring the exact lengths of the lunar month and the solar year.

I've said it before, but it cannot be said too often—we have been extremely lucky in being assigned a planet in exactly the location where ours is. Not only is Earth the only living planet that we know of, we have also been lucky in the choice of our one and only moon. It gives us tides, and with its soothing presence at night it can trigger quite lyrical feelings in people and other animals—but that's not what I want to talk about here.

One intriguing property of the Moon is that its diameter is 1/400 that of the Sun. This number wouldn't be special if it wasn't for the fact that the Sun is also 400 times farther away from us than the Moon. This implies that in our sky the Moon has exactly the same apparent size as the Sun and can cover it precisely. None of the three dozen other moons of our Solar System does that for inhabitants of their respective planets.

Only because the Moon has the same apparent size as the Sun, we are fortunate enough to be able to see the Sun's corona during total eclipses. If the Moon was smaller or farther away, the remaining brightness from the border of the solar disc would outglare the corona (this already happens in eclipses with the Moon at the most distant part of its orbit—so-called annular eclipses). If the Moon was bigger or closer, it would cover up the corona as well.

The spectacular sight of a total solar eclipse hits any given point on Earth once in four centuries on average. Thus, the peoples of ancient civilizations, who would know very little about what happened elsewhere, could get quite scared if they were confronted with such an unheard-of event. Nearly all cultures saw it as threatening. In the absence of any more plausible explanation, they assumed that a monster or dragon was eating the Sun. People made a lot of noise to chase away the monster and regain the Sun, and as it worked every time, they had little reason to doubt their beliefs. In

China, in particular, eclipses were predicted for the purpose of organizing the protest demonstration beforehand.

Cultures that were a bit more knowledgeable in astronomy, however, were quick to work out that all that was happening was that the path of the new moon crossed that of the Sun. However, one doesn't have to understand this to be able to detect the regularities in eclipses. A pinhole to project an image of the Sun (so one can observe partial eclipses as well) and a few centuries of patience are quite enough. Babylonian priests kept records of eclipses from about 3000 BC through to 300 BC, and over the centuries recognized all the repetitive patterns to be found in their occurrence. This goes to show that a full phenomenological description of such patterns, including prediction, is possible without the slightest hint of understanding the underlying mechanisms.

After Alexander the Great had taken Babylonia, the knowledge of eclipse patterns came to Greece. In 130 BC, the Greek astronomer Hipparchus used the data of a total eclipse to calculate the distance between the Earth and the Moon. He succeeded within 11 per cent. Hipparchus also used eclipse records to calculate the year length to be 365.2467 days, which is only about 10 parts per million off the value we believe in today (365.2422).

Through to the beginning of the twentieth century, eclipses have brought valuable insights for scientists. The first discovery of the spectral lines of the element helium were made during an eclipse (see FLASHBACK on p. 8). In 1919, the British astrophysicist Arthur Stanley Eddington (1882–1944) used a total eclipse for the verification of Einstein's prediction of how strongly the gravitational field of the Sun bends the path of starlight, and thus for the first time confirmed Einstein's General Theory of Relativity by experimental observation.

Since the beginning of the space age, eclipses are no longer of such particular value to science. If astronomers need to black out the Sun's glare, they can do so by including a black disc in satellite-based instruments like SOHO (see Chapter 1) rather than waiting for an eclipse. However, eclipses are still appreciated for their show qualities. The darkening which only lasts for a couple of minutes can lead to temperature drops of up to 10°C. If the weather is right, stars become visible. Wildlife can be fooled into thinking the Sun has set, most noticeable in that birds stop singing.

Thousands of fans travel to each eclipse, even if its path happens to swipe along a totally uninhabited stretch of the Pacific or the Sahara. In the wake of the eclipse of 11 August 1999, which was remarkable for choosing a path inhabited by hundreds of millions of people across Europe and the Near East, eclipse fever took hold of the whole Continent. I went to Germany to see it. As the weather was far from optimal, I stopped my travel at Heidelberg, on the edge of the totality zone, on the grounds that the eclipse would be behind clouds anyway. Just outside Heidelberg railway station I was lucky to catch a hole in the clouds at the right place and time to be able to see totality. There were no birds or stars, sadly, but a totally eclipsed Sun. And it is impressive. The next show will be 31 May 2003, when an annular eclipse is due to hit Scotland, Iceland, and the Shetlands.

Another fire within: early theories of vision

Except during sunrise, sunset, and the brief period of totality of an eclipse, it is dangerous to look directly at the Sun. The intense light can quickly burn the retina and thus lead to both pain and blindness. This simple observation enabled the Arabian scientist Alhazen (965–1040) to sweep away the Greek theories of vision that had dominated European thinking until then.

The Greek philosophers Empedocles (490– c.430 BC) and Plato (427–347 BC), and many of the later thinkers of the ancient world, assumed that seeing is an activity which is directed from the viewer towards the viewed object, a concept that has survived to this day in many expressions such as 'to look at something'. In the description by the Roman physician Galen (AD 129–99), which is strangely reminiscent of what we now know about bioluminescence, foodstuff is transformed in various steps until it produces the vision ray which supposedly emanates from the eye.

By pointing out the pain and damage the eye suffers from 'looking at' the Sun, Alhazen put the record straight and established that the only rays required for vision come from the object seen and penetrate the eye. He developed a fairly modern description of the eye and its function using the analogy with the camera obscura which he had studied in detail, effectively replacing Empedocles' romantic notion of the 'fire within' with the more prosaic dark chamber. It must have been the language barrier which robbed Alhazen of a more immediate impact in the western world—his fundamental work *Optics* was only translated into Latin a century and a half after his death.

Another Greek theory of vision, which Alhazen dismissed just as easily, had been put forward in the fifth century BC by the inventor of the concept of atoms, Democritus of Abdera, and his teacher Leucippus. Having very presciently postulated that the perception of smell requires invisibly small particles to travel from the object in question to the nose, they tried to develop an analogous description of vision. As they saw it, an atomically thin layer would peel off the object seen, retain the impression of its shape, and travel to the eye. While this theory had the directionality right, the atomists could not provide a plausible explanation as to why and how the shape envelope of a mountain could shrink so much that it could enter the eye.

Maybe the Greek philosophers would be pleased to know, however, that nowadays we can see with the help of rays fired at objects, and that we also have uses for atomically thin layers retracing the contours. Both these elements are found in the technique of electron microscopy, used to visualize objects smaller than the wavelength of visible light.

And the fire within also seems to be alive and kicking when we get into the kind of situation where we look into one another's eyes for extended periods of time and start seeing things in them which may not be there in a strictly Newtonian sense. What you see isn't always what you get, and nobody knew that better than one very famous poet.

Goethe's *Farbenlehre*

'More light' is what Germany's all-time favourite writer, Johann Wolfgang von Goethe (1749–1832) allegedly said before dying. Although it is far from clear whether these really were his last words, they would be very fitting, as a scientific interest in light and colour spanned half his lifetime and led to a series of three publications, his *Farbenlehre* ('Towards a theory of colour', first published in 1808–10). In repeated statements made to his assistant Johann Peter Eckermann towards the end of his long life, Goethe declared the *Farbenlehre* his most important work, although it was largely ridiculed by contemporaries as an embarrassing blunder of an otherwise brilliant poet and playwright.

Goethe's interest in light and colour perception was stimulated during his famous journeys to Italy, when he studied both colourful Renaissance paintings and sun-drenched southern landscapes. Well ahead of his time, he realized that colour perception was a phenomenon determined as much by the neurophysiology of the seeing organism as by the physical nature of the light involved. Unlike Newton, who emphasized the physical nature of colour by describing how each colour can be assigned to a particular wavelength or mixture of wavelengths, Goethe's interest was focused on the physiological aspect of colour. Half a century before Darwin, he pondered what we would now call the evolution of colour perception.

His major PR disaster, however, arose from his decision to attack Newton's optical theories, which were already the orthodoxy of his time, and present his own work as an alternative to Newton's, while it was really addressing the biological side of what Newton had described from a purely physical viewpoint. While Newton's followers dismissed optical illusions, afterimages, and the like as quirks that only distract from physical reality, Goethe regarded them as windows into the mind—a view that was revitalized by neuropsychology in the second half of the twentieth century.

In striking anticipation of Hering's blue–yellow axis (pp. 130–1), Goethe posited that all colours derive from two fundamental colours, yellow and blue, which in turn relate to white and black, respectively. One of his observations which seemed to confirm this was the following: an extremely bright light, e.g. the midday Sun, seen through a coloured filter will still appear white. At a somewhat smaller intensity, the bright light may appear as white with a hint of yellow, if the filter allows long wavelengths to pass, or with a hint of blue if it lets short wavelengths through.

Goethe's science was probably as sound as that of any other researcher, but came at a bad time. The greatest scientists of his time were essentially credited with making the step from subjective perception to objective observation. That subjective perception is in itself a subject worthy of scientific study was an idea whose time would only come much later.

Outlook

One of the reasons why I have dared to venture out into the rather unfamiliar territory of ancient and not so ancient history was that I wanted to remind you and myself that the scientific knowledge we have today about light and life is not the last word. Ideas and understanding change all the time. As science moves on, yesterday's heresy may become today's orthodoxy and then tomorrow's superstition. Although all the knowledge described in the first five chapters relies on sound scientific investigation, there is always the possibility that such an investigation has isolated the elephant's trunk and described it as a snake, or its foot as a pillar. As more integrated descriptions become feasible, future schoolchildren may find our misconceptions just as strange as Empedocles' vision rays, and they may marvel at our inability to understand dreaming or catch neutrinos.

The ability to make the most efficient use of light in technology (a topic which I left out of this book on purpose, focusing on the natural world) has set many milestones in the history of mankind. To this day technology involving better uses of light is often seen as intrinsically futuristic. This may have something to do with the visual appeal of laser swords and holograms in science fiction movies and TV series. But it is also true that the information capacity that can travel through light-guides grows even faster than that of electronics. Microengineers are already anticipating an electronic bottleneck situation, where those old-fashioned silicon chips will hold back the performance of networks mostly based on optical components.

As technology continues to learn from nature, using more vision-like information technology and more photosynthesis-like energy harvesting, the boundaries between technical and natural systems will begin to blur. Primitive retinas allowing blind patients (who lost their sight only in adulthood and thus have a fully developed visual cortex) to see at least a handful of pixels are already being tested.

Light is already one of the most important factors (if not the single most important factor) affecting life on Earth, and our lives as human beings in particular. If we learn to handle light as efficiently as nature does it, today's miracles will become tomorrow's every-day technology, adding a few more twists to the already closely entwined strands of Light and Life.

Further reading

CHAPTER 1

Lovelock, James (1995). *Gaia. A new look at life on Earth* (revised edn). Oxford University Press, Oxford.

The classic description of our living planet ...

Smil, Vaclav (2002). *The Earth's biosphere. Evolution, dynamics and change.* MIT Press, Cambridge, MA.

... and a more recent version of the story.

CHAPTER 2

Raven, Peter H., Evert, Ray F., and Eichhorn, Susan E. (1981). *Biology of plants* (6th edn). W. H. Freeman, New York.

A substantial textbook of plant biology.

Whatley, F. R. (1997). Changing views of photosynthesis. In: *Further milestones in biochemistry*, Vol. 3 (ed. M. G. Ord and L.A. Stocken). Elsevier, New York, pp. 23–65.

CHAPTER 3

Ganeri, Anita (1995). *Creatures that glow.* Victor Gollancz, London.

A book for young readers with amazing drawings and a glow-in-the-dark poster.

McElroy, William D. and Glass, Bentley (eds.) (1961). *A symposium on light and life.* Johns Hopkins University Press, Baltimore.

Probably out of print by now.

CHAPTER 4

Braitenberg, Valentino (1984). *Vehicles. Experiments in synthetic psychology.* MIT Press, Cambridge, MA.

Imaginative thinking, well presented.

Partonen, T. and Magnusson, A. (2001). *Seasonal affective disorder. Practice and research.* Oxford University Press, Oxford.

CHAPTER 5

Gregory, R. L. (1997). *Eye and brain. The psychology of seeing* (5th edn). Oxford University Press, Oxford.

Classic textbook that fits the pocket.

Turner, R. Steven (1994). *In the eye's mind. Vision and the Helmholtz–Hering controversy.* Princeton University Press, Princeton.

Very thorough but difficult-to-read account of this complex controversy.

CHAPTER 6

Assmann, Jan (1995). *Egyptian solar religion in the New Kingdom. Re, Amun and the crisis of polytheism.* Kegan Paul International, London.

Academic monograph translated from German.

Bortoft, Henri (1996). *The wholeness of nature. Goethe's way of science.* Lindisfarne Books, Herndon, VA.

Harrington, Philip S. (1997). *Eclipse! The what, where, when, why, and how guide to watching solar and lunar eclipses.* Wiley, New York.

Covers all eclipses through to 2017.

Hornung, Erik (1999). *Akhenaten and the religion of light.* Ithaca, Cornell University Press.

Translation of a popular account of Akhenaten's reign, originally published in German.

Singh, Madanjeet (1993). *The sun. Symbol of power and life.* Abrams,
New York.

Sumptuously illustrated whistle-stop tour of Sun religions around the globe.

Zajonc, Arthur (1993). *Catching the light.* Oxford University Press, Oxford.

A philosophical account of our relationship with light.

Index